FIRST CONCEPTS OF TOPOLOGY

THE GEOMETRY OF MAPPINGS OF SEGMENTS, CURVES, CIRCLES, AND DISKS

NEW MATHEMATICAL LIBRARY
published by
The Mathematical Association of America

Editorial Committee

Ivan Niven, Chairman (1981–83)   Anneli Lax, Editor
University of Oregon   *New York University*

W. G. Chinn (1980–82)   *City College of San Francisco*
Basil Gordon (1980–82)   *University of California, Los Angeles*
M. M. Schiffer (1979–81)   *Stanford University*

The New Mathematical Library (NML) was begun in 1961 by the School Mathematics Study Group to make available to high school students short expository books on various topics not usually covered in the high school syllabus. In a decade the NML matured into a steadily growing series of some twenty titles of interest not only to the originally intended audience, but to college students and teachers at all levels. Previously published by Random House and L. W. Singer, the NML became a publication series of the Mathematical Association of America (MAA) in 1975. Under the auspices of the MAA the NML will continue to grow and will remain dedicated to its original and expanded purposes.

FIRST CONCEPTS OF TOPOLOGY

THE GEOMETRY OF MAPPINGS OF SEGMENTS, CURVES, CIRCLES, AND DISKS

by

W. G. Chinn
San Francisco Public Schools

and

N. E. Steenrod
Princeton University

18

THE MATHEMATICAL ASSOCIATION
OF AMERICA

Illustrations by George H. Buehler

Eighth Printing

©Copyright 1966 by The Mathematical Association of America (Inc.)
All rights reserved under International and Pan-American Copyright
Conventions. Published in Washington, D.C. by
The Mathematical Association of America

Library of Congress Catalog Card Number: 66-20367

Complete Set ISBN-0-88385-600-X

Volume 18: ISBN-0-88385-618-2

Manufactured in the United States of America

Note to the Reader

This book is one of a series written by professional mathematicians in order to make some important mathematical ideas interesting and understandable to a large audience of high school students and laymen. Most of the volumes in the *New Mathematical Library* cover topics not usually included in the high school curriculum; they vary in difficulty, and, even within a single book, some parts require a greater degree of concentration than others. Thus, while the reader needs little technical knowledge to understand most of these books, he will have to make an intellectual effort.

If the reader has so far encountered mathematics only in classroom work, he should keep in mind that a book on mathematics cannot be read quickly. Nor must he expect to understand all parts of the book on first reading. He should feel free to skip complicated parts and return to them later; often an argument will be clarified by a subsequent remark. On the other hand, sections containing thoroughly familiar material may be read very quickly.

The best way to learn mathematics is to *do* mathematics, and each book includes problems, some of which may require considerable thought. The reader is urged to acquire the habit of reading with paper and pencil in hand; in this way mathematics will become increasingly meaningful to him.

The authors and editorial committee are interested in reactions to the books in this series and hope that readers will write to: Anneli Lax, Editor, New Mathematical Library, NEW YORK UNIVERSITY, THE COURANT INSTITUTE OF MATHEMATICAL SCIENCES, 251 Mercer Street, New York, N. Y. 10012.

The Editors

NEW MATHEMATICAL LIBRARY

1. Numbers: Rational and Irrational *by Ivan Niven*
2. What is Calculus About? *by W. W. Sawyer*
3. An Introduction to Inequalities *by E. F. Beckenbach and R. Bellman*
4. Geometric Inequalities *by N. D. Kazarinoff*
5. The Contest Problem Book I Annual High School Mathematics Examinations 1950–1960. Compiled and with solutions *by Charles T. Salkind*
6. The Lore of Large Numbers *by P. J. Davis*
7. Uses of Infinity *by Leo Zippin*
8. Geometric Transformations I *by I. M. Yaglom, translated by A. Shields*
9. Continued Fractions *by Carl D. Olds*
10. Replaced by NML-34
11. } Hungarian Problem Books I and II, Based on the Eötvös Competitions
12. } 1894–1905 and 1906–1928, *translated by E. Rapaport*
13. Episodes from the Early History of Mathematics *by A. Aaboe*
14. Groups and Their Graphs *by E. Grossman and W. Magnus*
15. The Mathematics of Choice *by Ivan Niven*
16. From Pythagoras to Einstein *by K. O. Friedrichs*
17. The Contest Problem Book II Annual High School Mathematics Examinations 1961–1965. Compiled and with solutions *by Charles T. Salkind*
18. First Concepts of Topology *by W. G. Chinn and N. E. Steenrod*
19. Geometry Revisited *by H. S. M. Coxeter and S. L. Greitzer*
20. Invitation to Number Theory *by Oystein Ore*
21. Geometric Transformations II *by I. M. Yaglom, translated by A. Shields*
22. Elementary Cryptanalysis—A Mathematical Approach *by A. Sinkov*
23. Ingenuity in Mathematics *by Ross Honsberger*
24. Geometric Transformations III *by I. M. Yaglom, translated by A. Shenitzer*
25. The Contest Problem Book III Annual High School Mathematics Examinations 1966–1972. Compiled and with solutions *by C. T. Salkind and J. M. Earl*
26. Mathematical Methods in Science *by George Pólya*
27. International Mathematical Olympiads—1959–1977. Compiled and with solutions *by S. L. Greitzer*
28. The Mathematics of Games and Gambling *by Edward W. Packel*
29. The Contest Problem Book IV Annual High School Mathematics Examinations 1973–1982. Compiled and with solutions *by R. A. Artino, A. M. Gaglione, and N. Shell*
30. The Role of Mathematics in Science *by M. M. Schiffer and L. Bowden*
31. International Mathematical Olympiads 1978–1985 and forty supplementary problems. Compiled and with solutions *by Murray S. Klamkin*
32. Riddles of the Sphinx *by Martin Gardner*
33. U.S.A. Mathematical Olympiads 1972–1986. Compiled and with solutions *by Murray S. Klamkin*
34. Graphs and Their Uses *by Oystein Ore.* Revised and updated *by Robin J. Wilson*
35. Exploring Mathematics with Your Computer *by Arthur Engel*
36. Game Theory and Strategy *by Philip D. Straffin, Jr.*
37. Episodes in Nineteenth and Twentieth Century Euclidean Geometry *by Ross Honsberger*
38. The Contest Problem Book V American High School Mathematics Examinations and American Invitational Mathematics Examinations 1983–1988. Compiled and augmented *by George Berzsenyi and Stephen B. Maurer*

Other titles in preparation.

Contents

Introduction		1
Part I	**Existence Theorems in Dimension 1**	5
	1. The first existence theorem	5
	2. Sets and functions	9
	3. Neighborhoods and continuity	16
	4. Open sets and closed sets	22
	5. The completeness of the real number system	31
	6. Compactness	37
	7. Connectedness	46
	8. Topological properties and topological equivalences	53
	9. A fixed point theorem	60
	10. Mappings of a circle into a line	62
	11. The pancake problems	64
	12. Zeros of polynomials	70
Part II	**Existence Theorems in Dimension 2**	75
	13. Mappings of a plane into itself	75
	14. The disk	79
	15. Initial attempts to formulate the main theorem	81
	16. Curves and closed curves	82
	17. Intuitive definition of winding number	84
	18. Statement of the main theorem	86
	19. When is an argument not a proof?	88
	20. The angle swept out by a curve	88
	21. Partitioning a curve into short curves	91
	22. The winding number $W(\varphi, y)$	94
	23. Properties of $A(\varphi, y)$ and $W(\varphi, y)$	98

24. Homotopies of curves	98
25. Constancy of the winding number	102
26. Proof of the main theorem	106
27. The circle winds once about each interior point	106
28. The fixed point property	108
29. Vector fields	110
30. The equivalence of vector fields and mappings	112
31. The index of a vector field around a closed curve	114
32. The mappings of a sphere into a plane	116
33. Dividing a ham sandwich	120
34. Vector fields tangent to a sphere	123
35. Complex numbers	127
36. Every polynomial has a zero	130
37. Epilogue: A brief glance at higher dimensional cases	133
Solutions for Exercises	137
Index	159

Introduction

Our purpose in writing this book is to show how topology arose, develop a few of its elements, and present some of its simpler applications.

Topology came to be recognized as a distinct area of mathematics about fifty years ago, and its major growth has taken place within the last thirty years. It is the most vigorous of the newer branches of mathematics and has been producing strong repercussions in most of the older branches. It got its start in response to the needs of analysis (the part of mathematics containing calculus and differential equations). However, topology is not a branch of analysis. Instead, it is a kind of geometry. It is not an advanced form of geometry such as projective or differential geometry, but rather a primitive, rudimentary form—one which underlies all geometries. A striking fact about topology is that its ideas have penetrated nearly all areas of mathematics. In most of these applications, topology supplies essential tools and concepts for proving certain basic propositions known as *existence* theorems.

Our presentation of the elements of topology will be centered around two existence theorems of analysis. The first, given in Part I, is fundamental in the calculus and was known long before topology was recognized as a subject. In working out its proof, we shall develop basic ideas of topology. This will show how topology got started, and why it is useful. Our second main theorem, given in Part II, is a generalization of the first from one to two dimensions. In contrast to the first, a topological concept is needed for its formulation. Its proof exhibits that peculiar blending of numerical precision and rough qualitative geometry so characteristic of topology. Both theorems have numerous applications. We shall present those having the strongest topological flavor.

The beginnings of topology can be found in the work of Karl Weierstrass during the 1860's in which he analyzed the concept of the *limit* of a function (as used in the calculus). In this endeavor, he reconstructed the real number system and revealed certain of its properties now called "topological". Then came Georg Cantor's bold development of the theory of point sets (1874–1895); it provided a foundation on which topology eventually built its own house. A second aspect of topology, called *combinatorial* or *algebraic* topology, was initiated in the 1890's by the remarkable work of Henri Poincaré dealing with the theory of integral calculus in higher dimensions. The first aspect, called *set-theoretic* topology, was placed on a firm foundation by F. Hausdorff and others during the period 1900–1910. A union of the combinatorial and set-theoretic aspects of topology was achieved first by L. E. J. Brouwer in his investigation (1908–1912) of the concept of dimension. The unified theory was given a solid development in the period 1915–1930 by J. W. Alexander, P. L. Alexandrov, S. Lefschetz and others. Until 1930 topology was called *analysis situs*. It was Lefschetz who first used and popularized the name *topology* by publishing a book with this title in 1930.

Since 1930 topology has been growing at an accelerated pace. To emphasize this point we shall mention a few of topology's achievements. It invaded the calculus of variations through the theory of critical points developed by M. Morse (Institute for Advanced Study, Princeton). It reinvigorated differential geometry through the work on fibre bundles by H. Whitney (Institute for Advanced Study, Princeton), the work on differential forms by G. de Rham (Lausanne), and the work on Lie groups by H. Hopf (Zürich). It enforced a minor revolution in modern algebra through the development of new foundations for algebra and a new branch called homological algebra. Much of this work is due to S. Eilenberg (Columbia University) and S. MacLane (University of Chicago). Topology gave a new lease on life to algebraic geometry via the theory of sheaves and cohomology, and it has found important applications to partial differential equations through the works of J. Leray (Paris) and M. Atiyah (Oxford).

Applications of topology have been made to sciences other than mathematics, but nearly all of these occur through some intervening mathematical subject. For example, the changes topology has made in differential geometry have initiated topological thinking in relativity theory. Topology has become a basic subject of mathematics, in fact, a necessity in many areas and a unifying force for nearly all of mathematics.

When a non-mathematician asks a topologist, "What is topology?", "What is it good for?", the latter is at a disadvantage because the questioner expects the kind of answer that can be given to analogous questions about trigonometry, namely, trigonometry deals with the determination of angles and is used to solve problems in surveying, navigation, and astronomy. The topologist cannot give such a direct answer; he can say, correctly, that topology is a kind of geometric thinking useful in many areas of advanced mathematics, but this does not satisfy the questioner who is after some of the flavor of the subject. The topologist can then bring out paper, scissors and paste, construct a Möbius band, and cut along its center line, or he can take some string and show how three separate loops can be enlaced without being linked. If he feels energetic, he can demonstrate how to take off his vest without removing his coat. These are parlor tricks, each based on a serious mathematical idea which would require at least several hours to explain. To present these tricks without adequate explanation is to present a caricature of topology.

To appreciate topology it is necessary to take the viewpoint of the mathematician and explore some of its successful applications. Most of these applications have in common that they occur *in the proof of an existence theorem*. An existence theorem is one which asserts that each of a certain broad class of problems has a solution of a particular kind. Such theorems are frequently the basic structure theorems of a subject. One of our principal aims is to demonstrate the power and flexibility of topology in proving existence theorems.

The existence theorem we shall prove in Part I answers the question: *When can an equation of the form $f(x) = y$ be solved for x in terms of y* ? Here $f(x)$ stands for a function or formula (such as $x^3 - \sqrt{1 + x^2}$) defined for real numbers x in some interval $[a, b]$ (such as $[2, 4]$), and y denotes a real number (such as $\frac{33}{2}$). The problem is: Does there exist a number x in the interval $[a, b]$ such that $f(x) = y$? Formulated for the example it becomes: Is there an x between 2 and 4 such that $x^3 - \sqrt{1 + x^2} = \frac{33}{2}$?

We emphasize that we are not asking for methods of finding the value or values of x in special cases. Instead we are seeking a broad criterion, applicable to each of many different problems, to determine whether or not a solution exists. Once the criterion assures us that a particular problem has a solution, we can start hunting for it with the knowledge that the search is not in vain.

The criterion given by our main theorem (stated in Section 1) requires the notion of the *continuity* of a function (defined in Section 3). The proof of the theorem (given in Sections 2–8) is based on two topological properties of the interval $[a, b]$ called *compactness* and *connectedness*. We give these concepts a thorough treatment because they are basic in modern mathematics.

The main theorem of Part II is an existence theorem which answers the question: *When can a pair of simultaneous equations $f(x, y) = a$ and $g(x, y) = b$ be solved for x and y in terms of a and b* ? A familiar example of such a problem is the pair of simultaneous linear equations

$$x - 2y = 3 \quad \text{and} \quad 3x + y = 5;$$

these can be solved readily by elimination. Here is a more difficult problem of the same type: Find a pair of numbers x, y satisfying the two equations

$$\frac{y \log x}{1 + x^2} = -\frac{1}{4} \quad \text{and} \quad x + 2y^3 = 10.$$

In particular, is there a solution such that x is between $\frac{1}{2}$ and 1, and y is between -1 and 2? Just as before, we are not asking for a list of methods for finding solutions (x, y), but rather for a test to determine whether or not there is a solution.

The criterion given by the second main theorem (stated in Section 18) requires the concept of the number of times a curve in a plane winds about a point of the plane. The proof also makes extensive use of the machinery of Part I having to do with compactness, connectedness, and continuity.

The applications of our main theorems are theorems concerning the existence of zeros of polynomials, fixed points of mappings, and singularities of vector fields.

In conclusion it may be well to say a few words about existence theorems in general. Their importance is granted immediately by mathematicians. Students, at first, may be somewhat skeptical. The reason is that there is quite a gap between the methods given in the proof of existence and the techniques the student must learn for finding solutions. The proof of existence must work in all cases however difficult; hence its methods tend to be complicated and tedious in application to particular cases. Most cases confronting the student are relatively simple, and therefore amenable to much simpler methods.

Consider for example the problem of finding zeros of polynomials. The equations presented first to the student are usually of low degree, have integer coefficients and can be factored by inspection. For less simple problems he learns to find the integral roots by testing the factors of the constant term. Then he learns a more complicated method for finding the rational roots. Finally, for rare and more desperate situations, he may learn a method of successive approximations due to Horner. The difficulty of acquiring these techniques is sufficient to make him forget the general question of the existence or non-existence of what he is seeking. If he is reminded of the question, he quickly relegates it to the domain of the metaphysical.

That it is not a metaphysical question becomes clear if we consider the history of the famous problems of trisecting an angle and squaring the circle using only straightedge and compasses. Since the time of Euclid, mathematicians and others have struggled with these problems, devising scheme on top of ingenious scheme. They invariably approached the problem with the tacit assumption that solutions exist. The problem was to find them. The amount of effort expended in this search must have been prodigious. It was not until the latter part of the nineteenth century that someone finally considered the possibility that solutions might not exist. Shortly thereafter, proofs of non-existence were forthcoming. Once the existence question was brought clearly to the fore, it was answered promptly. In modern research, existence questions come first; answers to them are absolutely vital in order that our theories have sound foundations.

PART I

Existence Theorems in Dimension 1

1. The first existence theorem

This section is devoted to the formulation of the main existence theorem of Part I. Its proof will be worked out in Sections 2–7 and summarized in Section 8. We shall lead up to its statement by examining a number of special cases. Recall that our problem is to formulate a criterion which will tell us in many cases whether or not an equation of the form $f(x) = y$ can be solved for x. To see what form the criterion might take, we examine cases where we know how to solve the equation completely.

Consider first the function $f(x)$ defined by the formula $x^2 + 1$ for x-values between -1 and $+2$. (The formula makes sense for x-values outside the interval -1 to 2, but we shall ignore this fact.) The function can be pictured from its graph shown in Fig. 1.1. The equation $y = x^2 + 1$ defines a parabola, and our graph is the piece of the parabolic curve between the vertical lines where $x = -1$ and $x = 2$.

Notice first that there is a lowest point on the curve at $x = 0$, $y = 1$. This can be restated precisely: $x^2 + 1$ is greater than or equal to 1 for all x between -1 and 2, and it has the minimum value 1 when $x = 0$. If we now look for the highest point on the curve for x between -1 and 2, we find that it occurs when $x = 2$ and $y = 5$, providing we interpret the phrase "x between -1 and 2" to include the end value $x = 2$ of the interval. If it were not included then there would be no highest point on the curve; because, no matter what point on the curve we take whose x-coordinate is less than 2, a higher point can be found by taking one whose x-coordinate is nearer to 2. To avoid such a situation, we shall include the end values -1 and 2. Then $x^2 + 1$ is less than or equal to 5 for all x such that $-1 \leq x \leq 2$, and the function has the maximum value 5 when $x = 2$.

Consider now the problem of starting with a y-value and trying to solve the equation $x^2 + 1 = y$ for a corresponding x-value in the interval -1 to 2. If the y-value exceeds the maximum 5, there is surely no solution. This is likewise the case if y is less than the minimum 1. However, if y is between 1 and 5, there is a solution. We can see this from the graph by drawing a horizontal line at a height above the x-axis equal to the y-value. If the line is too high or too low, it does not meet the curve. At a height between 2 and 5, it cuts the curve once, and between 1 and 2, it cuts twice. (A formula for x in terms of y is $x = \sqrt{y - 1}$.)

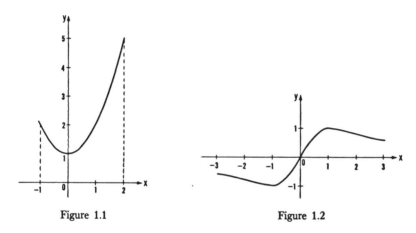

Figure 1.1   Figure 1.2

As a second example, let $f(x)$ be defined by the formula

$$\frac{2x}{x^2 + 1} \quad \text{for all } x\text{-values such that } -3 \leq x \leq 3.$$

Its graph is shown in Fig. 1.2. It is readily seen by inspecting the equation

$$y = \frac{2x}{x^2 + 1}$$

that a positive x-value gives a positive y-value, and a negative x-value gives a negative y-value. Moreover, changing the sign of the x-value changes only the sign of the y-value; hence the curve is symmetric with respect to the origin. The highest point on the curve occurs when x is 1 and then y is 1. To see that this is so, we do a bit of algebraic juggling of the difference $1 - f(x)$:

$$1 - f(x) = 1 - \frac{2x}{x^2+1} = \frac{x^2+1-2x}{x^2+1} = \frac{(x-1)^2}{x^2+1}.$$

Since this last expression can never be negative, it follows that $1 - f(x) \geq 0$, whence $f(x) \leq 1$. By symmetry, the lowest point on the curve occurs at $x = -1$ and $y = -1$. It is evident therefore that the equation $f(x) = y$ has no solution if $y > 1$ or if $y < -1$, but for each y-value such that $-1 \leq y \leq 1$, the equation can be solved. (Multiplying both sides of the equation by $x^2 + 1$, and solving the resulting quadratic gives $x = (1 - \sqrt{1 - y^2})/y$.)

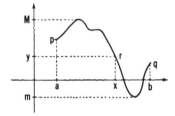

Figure 1.3 Figure 1.4

One might try to generalize from the two examples treated thus far and conjecture that, if $f(x)$ is any function defined for x-values such that $a \leq x \leq b$, then $f(x)$ has a maximum value M, a minimum value m, and for each y-value such that $m \leq y \leq M$ the equation $f(x) = y$ has a solution. Let us test this conjecture by picturing the graphs of several more functions, recalling that functions can be defined by specifying their graphs in any manner we please.

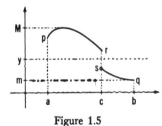 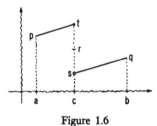

Figure 1.5 Figure 1.6

If the graph for $f(x)$ is a smooth curve as shown in Fig. 1.3, the conjecture appears to be true. A horizontal line at a height y between the heights m and M must intersect the curve. Even if the curve had some corners, as in Fig. 1.4, the conjecture still appears to be correct. However, if there is a break in the curve as in Fig. 1.5, the conjecture is seen to be false, because some horizontal lines pass through the break without meeting the graph. Functions whose graphs have such breaks do arise in mathematics in natural ways. They are called *discontinuous* functions.

The graph of such a function may not even have a highest or lowest point as in the example of Fig. 1.6 where, at the break in the graph at $x = c$, the point $(c, f(c))$ on the graph is at r.

We are prepared now to state the main theorem.

THEOREM. *If a function $f(x)$ is defined for all real numbers x in some closed interval $[a, b]$, if $f(x)$ has real numbers as values, and if it is continuous, then there is a minimum value m and a maximum value M of the function, and, for each y-value in the closed interval $[m, M]$, the equation $f(x) = y$ has at least one solution x in the interval $[a, b]$.*

The statement of the theorem is sometimes abbreviated thus: If the real-valued function $f(x)$ is defined and continuous for $a \leq x \leq b$, then it has a minimum value, a maximum value, and takes on all values between.

The expression "closed interval" means that the endpoints a, b are included as points of the interval, that is, the limitation on x is $a \leq x \leq b$. The expression "open interval" means that the endpoints are excluded. We shall denote the closed interval by $[a, b]$ and the open interval by (a, b). A "half-open" interval includes one endpoint but not the other, thus $(a, b]$ means $a < x \leq b$, and $[a, b)$ means $a \leq x < b$.

How does the theorem help us to decide whether or not we can solve $f(x) = y$ in the case of a particular function $f(x)$, known to be continuous, and a particular y-value? If we can determine the minimum m and maximum M of $f(x)$, we have only to ask if $m \leq y \leq M$. In many cases it may be difficult to find m and M. However, it is usually easy to compute a number of values of the function. If, for some x-value c, we observe that $f(c) < y$, and for another d, that $y < f(d)$, then the theorem asserts that there is an x in $[c, d]$ (or $[d, c]$ if $d < c$) such that $f(x) = y$. For example, if $f(x)$ is $x^3 - \sqrt{1 + 4x}$, we have $f(0) = -1$, and $f(2) = 5$. Therefore $x^3 - \sqrt{1 + 4x} = 2$ has a solution in the interval $[0, 2]$.

We must emphasize that the importance of the theorem lies in its generality. It tells us what we can always count on finding in a great variety of circumstances. In numerous special cases like $x^2 + 1$, it is of no use to us because the facts we are after are readily available. The theorem shows its power as soon as it is applied to complicated functions. But more important, it is the first theorem of a general theory of continuous functions.

It must also be emphasized that our formulation of the theorem is incomplete. We have not defined precisely what is meant by the continuity of f ; we have given an intuitive description based on geometric pictures —the graph of f is a curve with no breaks—but this is only a substitution of one undefined term for another. The next two sections lead up to a precise definition of continuity.

Exercises

1. Find the minimum value and maximum value of $f(x) = 4 + 2x - x^2$ in the interval $0 \leq x \leq 3$. For what values of y are there no corresponding values of x in the interval? For what values of y is there only one value of x in the interval? two values of x in the interval?

2. The function $f(x) = x^3 - 5$ takes on the value -4 at $x = 1$ and the value 3 at $x = 2$; it is continuous in the interval $1 \leq x \leq 2$. How does the theorem imply that $\sqrt[3]{5}$ has a value between 1 and 2?

3. Between what two positive integral values of x is there a zero for the polynomial $x^2 - 2x - 4$?

4. Find the minimum value and maximum value for $f(x) = 1/x$ in the interval $0 < x \leq 5$.

5. Find the minimum value and maximum value for $f(x) = 3$ in the interval $0 \leq x \leq 7$.

6. Find the minimum value and maximum value for $f(x) = x$ in the interval $0 \leq x < 5$.

2. Sets and functions

Throughout this book, we shall be primarily interested in geometric configurations. These are subsets of a euclidean line, plane, or space. For convenience, we shall assume that cartesian coordinates have been introduced into the line, plane, or space so that each point x is specified by its coordinates which form an ordered set of n real numbers $(x_1, x_2, \cdots, x_n)$, that is, an n-tuple of real numbers. Thus n is 2 in the case of the plane, n is 3 for three-dimensional space, and $n = 1$ for the line. The set of all real numbers will be denoted by R and the set of all n-tuples of real numbers by R^n. We shall think of these geometrically; thus $R = R^1$ is the number line (a line with a coordinate system), R^2 denotes a plane with a coordinate system, and R^3 denotes a three dimensional space with a coordinate system.

From the program set forth in both the Introduction and in the preceding section, it can be seen that considerable emphasis will be placed on the study of functions. In this book, we shall use the word "function" in a sense which is somewhat broader than is customary in the more elementary courses up through the calculus, and we shall indicate in this section the scope of generality we have in mind. In our development, we shall use the customary language and notation of set theory. Not many terms from this vocabulary will be used, but those that we do use will be used quite frequently. We present here the terms and notation needed in what follows.

If X is a set, then $x \in X$ means that x is an element of the set X. If A and X are sets, then $A \subset X$ (read as: A is contained in X) means that each element of A is an element of X, and A is called a *subset* of X. In most cases, we shall be dealing with subsets of a line, a plane, or space (that is, $X \subset R^n$ for some positive integer n), and therefore we shall often refer to the elements of X as points. If $A \subset X$ and $B \subset X$, then their *intersection* $A \cap B$ consists of all points common to A and B, and their *union* $A \cup B$ consists of all points either in A or in B or both. The empty set is denoted by $\emptyset$, and it is contained in any set. Thus $A \cap B = \emptyset$ means that A and B have no point in common. If $A \subset X$, then the *complement* of A in X, denoted by $X - A$, consists of all points of X not in A.

In advanced mathematics, the word function is used in an extremely broad sense; in fact, it appears as a fundamental concept of all mathematics. The following definition accords with this broadest usage.

DEFINITION. A function f consists of three things: a set X called the *domain* of f, a set Y called the *range* of f, and a rule which assigns to each element of X a corresponding element of Y. The notation $f: X \to Y$ is to be read: f is a function with domain X and range Y; or, briefly, f is a function from X to Y. If $x \in X$, the statement "f assigns to x the element $y \in Y$" is abbreviated $y = fx$ (in keeping with current practice, we omit the parentheses from $f(x)$). In case Y is exactly the set of all values fx for $x \in X$, we say that f is a function from X onto Y.

In the calculus, for example, a function usually means a real-valued function of a real variable; that is, its domain and range are subsets of R. Moreover, it is frequently supposed that the function is given by some formula such as $\sqrt{1-x}$. In such a case, it is customary not to describe either the domain or the range. It is tacitly assumed that the domain X is the set of those real numbers for which the formula makes sense (for example, $\sqrt{1-x}$ is defined for all $x \leq 1$ including negative numbers). The range Y is often taken to be exactly the set of all values of the function (for example, for $\sqrt{1-x}$, it is the set of all $y \geq 0$). Functions of this kind will be called *numerical* functions.

In more advanced work, it cannot be supposed that f is given by a formula from which the domain X and range Y can be worked out. Moreover, we do not wish X and Y to be restricted to subsets of R. In this book, X and Y will usually be subsets of euclidean spaces of possibly different dimensions: $X \subset R^m$ and $Y \subset R^n$. So we must be careful to specify X and Y for each function considered. Also, we shall not always suppose that Y is exactly the set of values of the function;

it may be larger. Let us consider first some familiar examples of geometrically defined functions.

A *translation* of the plane is a function $f: R^2 \to R^2$. It is the result of a rigid and uniform motion in which each point traverses a line segment or vector; the vectors are the paths of the various points, are all parallel, and have the same length and direction. A translation is specified by the vector for just one point because the others can be constructed from it. Thus, if we know that f carries the point p into the point q, then it will carry a point p' into the point q' such that p, q, p', q' form a parallelogram. For example, if f carries the origin $(0, 0)$ into the point $(2, -3)$, then it carries (x_1, x_2) into the point $(x_1 + 2, x_2 - 3)$. Therefore f is given by the formulas $y_1 = x_1 + 2$ and $y_2 = x_2 - 3$.

A *rotation* of the plane is a function $f: R^2 \to R^2$, again resulting from a rigid motion, this time about a fixed point z called the center of the rotation. Each circle with center z is carried by f onto itself; and each ray (half-line including initial point) issuing from z is carried onto another ray. The angle formed by these two rays is called the angle of rotation, and its measure in degrees does not depend on the initial ray. The rotation is specified by its center and angle of rotation.

A *reflection* in a line L of R^2 is a function $f: R^2 \to R^2$; it is a rigid motion that leaves fixed each point of L and interchanges the two sides of L. It is most easily visualized as the result of the rotation in space of the plane about the line L through $180°$.

It can be shown that the result of any rigid motion of the plane onto itself is a translation, a rotation, a reflection, or a reflection followed by a translation. The shapes and sizes of the configurations in the plane are not altered; only their positions and orientations may be changed. These functions are the *congruences* of elementary geometry.

A *similarity* of the plane is a function $f: R^2 \to R^2$ which alters all lengths by the same factor r. As an example, choose a point z of R^2, set $fz = z$, and, for each other point $x \neq z$, define fx to be the endpoint of the segment (or vector) issuing from z which has the same direction and twice the length of the segment from z to x. This f alters all lengths by the factor $r = 2$. Such an f, with $r > 1$, is called an *expansion* centered at z. When $r < 1$, f is called a *contraction*. A similarity with $r = 1$ is one of the rigid motions described above. A similarity with $r \neq 1$ always has a fixed point z, and it is the result of a contraction or expansion with center z followed by a rotation about z or a reflection in some line through z. A similarity always carries straight lines into straight lines, and it does not change the measures of angles between lines. It can alter the size, location, and orientation of a configuration, but it does not alter its shape.

EXISTENCE THEOREMS IN DIMENSION 1

Let L be a line in the plane R^2. The *perpendicular projection* $f: R^2 \to L$ assigns to each point x of R^2 the foot fx of the perpendicular from x to L.

Let S be the surface of a sphere in R^3 with center z. The *radial projection* $f: R^3 - z \to S$ assigns to each point x of R^3 different from z the point fx where the ray from z through x intersects S.

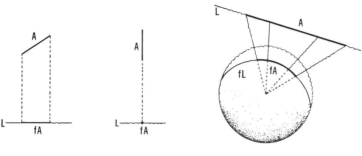

Figure 2.1 Figure 2.2 Figure 2.3

The foregoing examples indicate in part the kind of functions that will interest us. In order to picture such functions and to make significant statements about them, one uses the notions of *images* and *inverse images*. If $f: X \to Y$ and $A \subset X$, then the *image* of A under f, denoted by fA, is the subset of Y of values fx for all $x \in A$. Precisely, to say that a point $y \in Y$ is in fA means that there is at least one $x \in A$ such that $fx = y$. One can think of fA as the result of applying f to all of A. For example, under a rigid motion or a similarity $f: R^2 \to R^2$, any straight line L of R^2 has as its image a straight line fL of R^2. Under perpendicular projection $f: R^2 \to L$, each line segment A of R^2 has as its image fA a line segment of L (see Fig. 2.1), or a single point of L in case A is perpendicular to L (Fig. 2.2). Under radial projection $f: R^3 - z \to S$, each straight line L of R^3 not passing through z has as its image fL a great semi-circle of S with endpoints excluded (see Fig. 2.3).

The origin of our use of the word "image" in this sense is evident if we consider any reflection of R^3 in a plane.

If $f: X \to Y$ and $B \subset Y$, then the *inverse image* of B under f, denoted by $f^{-1}B$, is the subset of X consisting of points x such that $fx \in B$. Under perpendicular projection $R^2 \to L$, the inverse image of a single point y of L is the line perpendicular to L at y, and the inverse image of a line segment is the strip between the two lines perpendicular to L at the ends of the segment. Under radial projection $R^3 - z \to S$, the inverse image of a point y of S is the ray from z through y with z deleted. The inverse image of a circular region on S is a solid cone with the vertex deleted.

Let $f: X \to Y$ and let A be a subset of X. Then the image of A is also in Y. If B is a subset of Y such that B contains fA, then the function $g: A \to B$ defined by $gx = fx$ for all $x \in A$ is called the *restriction* of f to A and B. Most frequently we shall need to restrict only the domain of f, and, in this case, the restriction of f to A and Y is denoted by $f \mid A$ (read as: f restricted to A). For example, if $f: R^2 \to R^2$ is a translation, and L is a line in R^2, then $f \mid L$ displaces L to a parallel line.

If we have two functions $f: X \to Y$ and $g: Y \to Z$, then we can *compose* the two functions to form a new function denoted by $gf: X \to Z$; it attaches to each x in X the element $g(fx)$ in Z. For example, let f and g be translations $R^2 \to R^2$ where f moves each point 2 units toward the east, and g moves each point 2 units to the north. Then gf is the translation moving each point $2\sqrt{2}$ units to the northeast. As a second example, let f and g be numerical functions $R \to R$ given by the formulas

$$fx = x^2 + 1, \qquad gx = 2 - x.$$

Then the compositions gf and fg can be formed, and they are given by the formulas

$$gfx = g(fx) = 2 - (x^2 + 1) = 1 - x^2$$

$$fgx = f(gx) = (2 - x)^2 + 1 = 5 - 4x + x^2.$$

Some simple functions are so inconspicuous that one must be reminded of their presence. We call attention first to the *constant* functions: a function $f: X \to Y$ is a constant function if the image fX is a single point of Y. There is one constant function for each point of Y. Next, we mention the *identity* functions: for each set X, the function $f: X \to X$ such that $fx = x$ for every $x \in X$ is called the *identity* function of X. Finally, if $A \subset X$, the function $f: A \to X$ such that $fx = x$ for every $x \in A$ is called the *inclusion* function. Clearly, the inclusion is the restriction of the identity function to the subset. Any restriction of a constant function is constant.

A function $f: X \to Y$ is called *one-to-one* (abbreviated: 1–1) if each point of Y is the image under f of one and only one point of X.† In this case, the function which assigns to each point y in Y the unique point x in X such that $fx = y$ is called the *inverse* function of f and

---

† Some authors say that $f: X \to Y$ is one-to-one if each point of Y is the image of at most one point of X, allowing thereby that fX may not be all of Y; and in case $fX = Y$, they say that f is one-to-one and onto.

is denoted by f^{-1}: $Y \to X$. For example, every rigid motion of the plane is 1-1. If f is the translation given by a line segment from p to q, then f^{-1} is the translation given by a line segment from q to p. The inverse of a rotation is a rotation with the same center and an angle equal in magnitude but opposite in sign. A similarity of the plane is 1-1. The inverse of an expansion by a factor $r > 1$ is a contraction by the factor $1/r$ and it has the same center.

In dealing with numerical functions, one tries to obtain the formula for the inverse function by solving $y = fx$ for x in terms of y. Thus the square root is the inverse of the square, and the logarithm is the inverse of the exponential. In case the inverse images of certain points are not single points, one takes a restriction of the function f to subsets so as to obtain a 1-1 function. For example, if $fx = x^2$, from $y = x^2$ one obtains $x = \sqrt{y}$ and $x = -\sqrt{y}$, but if we restrict the domain of f to the subset A of positive numbers and zero, and restrict its range to the same set, then this restricted function g is 1-1, and its inverse is the usual square root function $g^{-1}y = \sqrt{y}$. In the case of the exponential function $fx = 10^x$, we need to restrict only the range of f to the set of positive numbers to obtain a 1-1 function.

Although our examples of functions have been quite varied, they still fail to suggest how broad and basic is the concept of function. As an example of a function in the broad sense, consider the concept "the mother of the boy". The domain X is the set of boys, the range Y is the set of women, and to each boy x is assigned the woman fx who is his mother. Such examples abound in our experiences: the color of a book, the roof of a house, etc. Whenever the word "of" is used, there is a function in the offing. This is equally so when the possessive form of a noun appears; for example, the boy's mother.

Functions are omnipresent in science. The outcome of a chemical reaction is a function of the reagents brought together. The outcome of a physical experiment is a function of the conditions set up by the experimenter.

Coming back to mathematics, there are examples of expressions of the form "the this of that" occurring everywhere: the area of a circle, the midpoint of a line segment, the bisector of an angle, the union of two sets, etc. Each of these is a function. In the case of the union of two sets, an element of the domain of the function is a *pair* (A, B) of subsets of a given set X, and its range is the set of subsets of X.

Many functions can be pictured geometrically. For example, the sum of two numbers $x + y$ is a function $f: R^2 \to R^1$ which can be visualized as a projection of a plane into a line. Regard each instance of a pair (x, y) as a point in R^2. The line whose equation is $x + y = 3$ is $f^{-1}3$. The inverse images of other numbers form the family of parallels (see Fig. 2.4). If we picture the range R^1 as a line cutting through this family of parallels at right angles, then f can be regarded as the perpendicular projection onto this line.

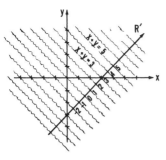

Figure 2.4   Figure 2.5

Exercises

1. If A, B, C are subsets of a set X, show that
$$(A \cup B) \cap C = (A \cap C) \cup (B \cap C)$$
and illustrate the result with a diagram of sets in R^2.

2. If A and B are subsets of X, show that
$$(X - A) \cap (X - B) = X - (A \cup B).$$

3. If A and B are subsets of a set X, and if $A \subset B$, show that
$$A \cap (X - B) = \varnothing.$$

4. Prove the following properties of images under $f: X \to Y$:
   (a) If A and B are any subsets of X, then $f(A \cup B) = fA \cup fB$,
   (b) $f(A \cap B) \subset fA \cap fB$,
   (c) if $A \subset B \subset X$, then $fA \subset fB$.

5. Stereographic projection from the north pole is a function f in which the domain X is the sphere with the north pole omitted and the range Y is a plane parallel to the equator but not containing the north pole. f assigns to each point x of the sphere (excepting the north pole) the point y where the ray from the north pole through x meets the plane (see Fig. 2.5). Suppose that the plane Y is tangent to the sphere at its south pole S.
   (a) Describe the image of a parallel of latitude.
   (b) Describe the image of a meridian circle.
   (c) What is the inverse image of a line segment in Y ?

(d) The image of the equator is a circle on the plane. How is the radius of the image circle related to the radius of the sphere?

(e) What is the inverse image of a ray through the south pole?

6. If f is the translation $R^2 \to R^2$ given by the formulas $y_1 = x_1 + 3$, $y_2 = x_2 - 4$, and g is the reflection given by $y_1 = -x_1$ and $y_2 = x_2$, find the formulas for the compositions gf, fg, for the inverses f^{-1}, g^{-1}, $(gf)^{-1}$, and the compositions $f^{-1}g^{-1}$ and $g^{-1}f^{-1}$. Is $(gf)^{-1} = g^{-1}f^{-1}$?

7. For the following, let $f: X \to Y$.

(a) If A, B are subsets of Y, prove that

$$f^{-1}(A \cup B) = f^{-1}A \cup f^{-1}B \quad \text{and} \quad f^{-1}(A \cap B) = f^{-1}A \cap f^{-1}B.$$

(b) Find $f^{-1}Y$ and $f^{-1}\emptyset$.

(c) If $A \subset B$, how are the inverses $f^{-1}A$ and $f^{-1}B$ related?

8. If $f: X \to Y$ and $g: Y \to Z$, show that $(gf)^{-1}C = f^{-1}(g^{-1}C)$ for each $C \subset Z$. If f and g are 1—1, show that gf is 1—1, and $(gf)^{-1} = f^{-1}g^{-1}$.

3. Neighborhoods and continuity

If x and y are two points of R^n, the *distance* from x to y means the ordinary straight line distance. It is denoted by $d(x, y)$, and in terms of the coordinates of x and y, it can be computed from the following formula based on a generalization of the theorem of Pythagoras:

$$d(x, y) = \sqrt{(x_1 - y_1)^2 + (x_2 - y_2)^2 + \cdots + (x_n - y_n)^2}.$$

In case $n = 1$, the formula reduces to $d(x, y) = |x - y|$ (the absolute value of $x - y$). Actually we shall not use this formula directly. Instead, we shall use only certain properties of the distance function which can be proved using the formula. These are well-known properties, and we list them now without proof.

First, if x and y are different points then $d(x, y) > 0$. Secondly, $d(x, x) = 0$ for all x. Next, for all pairs of points x, y, the distance is symmetric: $d(x, y) = d(y, x)$. Finally, for any three points x, y, z, we have

$$d(x, z) \leq d(x, y) + d(y, z).$$

This last is called the "triangle inequality" because it asserts that the sum of the lengths of two sides of a triangle exceeds the length of the third side.

§3] NEIGHBORHOODS, CONTINUITY

Recall that in our preliminary examination of various graphs in Section 1, whether or not the graph had any breaks was vital to our conclusion; it is for this reason that the main theorem specifies that the function must be continuous. The intuitive description of continuity was based on geometric pictures; the precise definition will be given in terms of another concept that we shall now define: the concept of a neighborhood.

DEFINITION. Let X be a subset of R^n, let x be a point of X, and let r be a positive real number. Then we define the *neighborhood* of x in X of radius r to be the set of all points of X whose distance from x is less than r. The neighborhood is denoted by $N(x, r, X)$, and this is abbreviated to $N(x, r)$ whenever the X is understood.

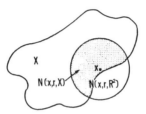

Figure 3.1

For example, if $X = R^n$ and $n = 2$, then $N(x, r, R^2)$ is just the interior of a circle with center x and radius r. Similarly the neighborhood of x in R^3 of radius r, $N(x, r, R^3)$, is the interior of a sphere with center x and radius r. In case $n = 1$, then $N(x, r, R)$ is just the open interval $(x - r, x + r)$ with midpoint x and length $2r$. Whenever X is not all of R^n, then $N(x, r, X)$ is just that part of X which lies inside $N(x, r, R^n)$ (see Fig. 3.1). It is the intersection of X with the neighborhood in R^n:

$$N(x, r, X) = X \cap N(x, r, R^n).$$

We come now to the important concept of the continuity of a function.

DEFINITION. Let $f: X \to Y$ be a function such that $X \subset R^m$ and $Y \subset R^n$ and let $x \in X$. We shall say that f *is continuous at* x if, for each neighborhood of fx in Y, there is some neighborhood of x in X whose image under f lies in the neighborhood of fx under consideration.

To express this condition briefly, we follow the customary notation of the calculus and denote the radii of these neighborhoods by ϵ and δ. Then the requirement can be restated: for each positive number ϵ, there is a positive number δ such that

$$fN(x, \delta, X) \subset N(fx, \epsilon, Y).$$

We shall say that f is *continuous* if it is continuous at every point of X.

If we interpret δ and ϵ as measures of nearness, then the definition may be paraphrased: one can make fx' be near to fx by requiring x' to be sufficiently near to x. An even rougher paraphrase is: a small change in x produces a small change in fx.

In topology we are primarily interested in continuous functions, and we abbreviate the expression "continuous function" by the word "mapping" which in turn is often shortened into "map". Thus stereographic projection is a *mapping* of the surface of a sphere with the pole deleted onto a plane. However, in order to illustrate what the definition of continuity means, let us examine several examples of discontinuous functions.

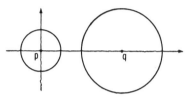

Figure 3.2

As a first example, let $f: R^2 \to R^2$ leave fixed each point of the plane except for a single point, say p, and let $fp = q$ be some other point. To be specific, we can take p to be the origin of coordinates $(0, 0)$ and q to be $(1, 0)$. Then f is continuous at all points except p. To see that it is not continuous at p, take ϵ to be half of $d(p, q)$, so that $N(q, \epsilon) = N(q, \frac{1}{2})$ is the interior of a circle of radius $\frac{1}{2}$ about $(1, 0)$ (see Fig. 3.2). Then no neighborhood of p is mapped into $N(q, \frac{1}{2})$, for each neighborhood of p contains points not in $N(q, \frac{1}{2})$, and since these are left fixed by f, their images are not in $N(q, \frac{1}{2})$. (Fig. 3.2 shows the neighborhood $N(p, \frac{1}{4})$; among its points only p has its image fp in $N(q, \frac{1}{2})$). Since for $\epsilon = \frac{1}{2}$ there is no corresponding $\delta > 0$ such that $fN(p, \delta) \subset N(q, \frac{1}{2})$, f is not continuous at p. The intuitive geometric picture is that f rips the point p out of the plane, and then pastes it down on q.

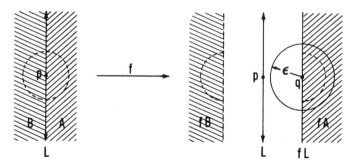

Figure 3.3

§3] NEIGHBORHOODS, CONTINUITY 19

Our second example displays a slightly different type of discontinuity. Divide the plane R^2 into two half-planes A, B, where A includes all points (x_1, x_2) such that $x_1 \geq 0$, and B is its complement. Notice that the vertical line L where $x_1 = 0$ is included in A (see Fig. 3.3). Define f to be a function from R^2 to R^2 so that the restriction of f to A is the translation to the right by one unit, and $f \mid B$ is the translation to the left by one unit. The two halves are separated, one moving to the right, the other to the left. Since $L \subset A$, the line L moves to the right. The function is continuous at each point not on L, and is discontinuous at each point of L. To prove the latter statement, let $p \in L$, let $q = fp$, and let ϵ be any positive number less than 2. Then the neighborhood of q in R^2 of radius ϵ, namely $N(q, \epsilon)$, contains no point of fB, and each $N(p, r)$ contains points of B. As shown in Fig. 3.3, the left half of any neighborhood of p (indicated by a dotted circle) is moved by f away from $N(q, \epsilon)$. Hence there is no corresponding δ such that the image of the neighborhood of p in R^2 of radius δ is contained in the neighborhood of q of radius ϵ, that is, such that

$$fN(p, \delta) \subset N(q, \epsilon).$$

The intuitive picture is that f rips the plane apart along L and separates the two halves.

To see that the definition agrees with our intuitive notion, we can test it against the examples of discontinuity for the graphs depicted in Section 1. In Fig. 1.5, if the radius ϵ of the neighborhood $N(r, \epsilon)$ is less than half the distance $d(r, s)$, then any neighborhood of c includes points of X whose images do not lie in this ϵ-neighborhood of $fc = r$. Similarly, in Fig. 1.6, if ϵ is less than the smaller of the distances $d(r, s)$ and $d(r, t)$, then any neighborhood of c again includes points whose images do not lie in the neighborhood of $fc = r$ of radius ϵ.

To prove that a function has a discontinuity at some point, it is only necessary to display a single $\epsilon > 0$ for which no δ exists. To prove continuity is often more difficult because we must show how to find a number δ corresponding to each possible choice of ϵ (that is, we must display δ as a function of ϵ). However, there are a number of simple functions where this is not difficult, and we consider these now.

For any $X \subset R^n$, the identity function $f: X \to X$ is continuous. Recall that $fx = x$ for all $x \in X$. Corresponding to $x \in X$ and an $\epsilon > 0$, take $\delta = \epsilon$. Then it is clear that f maps $N(x, \delta, X)$ into $N(fx, \epsilon, X)$, for these two neighborhoods coincide since f leaves all points fixed. Similarly, if $A \subset X$, then the inclusion function $f: A \to X$ is continuous; hence it is a mapping. Again we take $\delta = \epsilon$, and we use the fact that $N(x, \delta, A) = A \cap N(x, \delta, X)$.

Any constant function $f: X \to Y$ is continuous. In this case fX is a single point, say q, of Y; hence $fN(x, r, X) \subset N(q, \epsilon, Y)$ for every x, and for every $\epsilon > 0$, and $r > 0$, so we may take $\delta = \epsilon$, for example.

Any rigid function $f: X \to Y$ is continuous. By "rigid" we mean a function which does not alter distances:

$$d(fx, fx') = d(x, x') \quad \text{for all } x, x' \in X.$$

For example, the translations, rotations, and reflections of R^2 are rigid. To prove continuity at $x \in X$, we take $\delta = \epsilon$ for each $\epsilon > 0$. Then, if $x' \in N(x, \delta, X)$, we have $d(x, x') < \epsilon$, therefore $d(fx, fx') < \epsilon$, hence $fx' \in N(fx, \epsilon, Y)$. In words, the ϵ-neighborhood of x is carried by f into the ϵ-neighborhood of fx because of the rigidity of f.

Any function $f: X \to Y$ that shrinks or contracts all distances is continuous. The requirement is expressed by

$$d(fx, fx') \leq d(x, x') \quad \text{for all } x, x' \in X.$$

Again we take $\delta = \epsilon$ for all x, and we apply the argument of the preceding paragraph.

Any similarity function $f: X \to Y$ is continuous. The condition here is that all distances should be changed by the same common factor, say k:

$$d(fx, fx') = k\, d(x, x') \quad \text{for all } x, x' \in X.$$

If $0 \leq k \leq 1$, then f is a contraction and the preceding paragraph applies. When $k > 1$ we take $\delta = \epsilon/k$ for all points x. Then $x' \in N(x, \epsilon/k, X)$ implies $d(x, x') < \epsilon/k$. This can be written $k\, d(x, x') < \epsilon$. The equality above yields $d(fx, fx') < \epsilon$, hence $fx' \in N(fx, \epsilon, Y)$. For example, if $k = 2$ and f doubles distances, then the neighborhood of x of half the radius of the neighborhood $N(fx, \epsilon, Y)$ is carried by f into the latter.

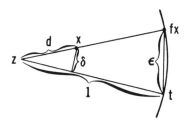

Figure 3.4

The radial projection onto the surface of a sphere S from its center z is continuous as a function $f: R^3 - z \to S$. In projecting two points in the exterior of S down onto S, f contracts the distance between them, so it is obvious that f restricted to the exterior of S is continuous. However, f restricted to the interior expands distances; the projection onto S expands distances between pairs of points more and more as they move inside toward the center z. To prove continuity at a point $x \neq z$, the expression for δ as a function of ϵ is rather involved, but can nevertheless be obtained with a bit of algebra by considering various similar triangles (see Exercise 9 below). Geometrically, we can obtain a δ thus: Let t be a point of intersection of S and the sphere with center fx and radius ϵ. The cross-section on the plane through the three points z, fx, and t is shown in Fig. 3.4. Let δ be the perpendicular distance from x to the line zt. Then each point of the interior of the sphere $N(x, \delta)$ projects into $N(fx, \epsilon, S)$.

Exercises

1. For which triples of points x, y, z will equality hold in the triangle inequality; that is,
$$d(x, z) = d(x, y) + d(y, z)?$$

2. If x' is a point of the neighborhood $N(x, r, X)$, show that there is an $r' > 0$ such that
$$N(x', r', X) \subset N(x, r, X).$$
What is the largest value of r' which assures this?

3. If, in testing the continuity of a function $f: X \to Y$ at a point x, a number $\delta > 0$ has been found which does for $\epsilon = \frac{1}{2}$, why will the same δ do for all $\epsilon \geq \frac{1}{2}$?

4. If, as in Exercise 3, a $\delta > 0$ has been found which does for $\epsilon = \frac{1}{2}$, why will any smaller value of δ do for $\epsilon = \frac{1}{2}$?

5. Divide the plane R^2 into two parts A and B where A consists of all points inside and on a circle C with center z and radius 1, and B is the complement of A in R^2. Define $f: R^2 \to R^2$ by $f | A$ rotates A about its center through an angle of $90°$, and f leaves fixed each point of B. Where is f continuous and where is it discontinuous? At a point of discontinuity, for what values of ϵ are there no corresponding δ's?

6. If L is a line or a plane in R^3, state why the perpendicular projection $f: R^3 \to L$ is continuous.

7. Let S be the spherical surface of radius 1 with center at the origin of R^3, let $p = (0, 0, 1)$ be the north pole of S, and let $f: S - p \to R^2$ be the stereographic projection from p of $S - p$ onto the equatorial plane. Construct a diagram which shows that f is continuous. Over what part of $S - p$ is f a contracting function? Show that f is 1–1, and that $f^{-1}: R^2 \to S - p$ is also continuous. What is the image under f of the deleted neighborhood $N(p, r, S) - p$ for r-values less than 1? Why is it impossible to define fp so that the extended function is continuous?

8. If f and g are continuous functions from an interval $[a, b]$ to R, show that $f + g$ and $f - g$ are continuous. Hint: To demonstrate the continuity at x of $hx = fx + gx$ and of $kx = fx - gx$, estimate $|hx - hx'|$ and $|kx - kx'|$ with the help of the triangle inequality

$$|(fx \pm gx) - (fx' \pm gx')| \leq |fx - fx'| + |gx - gx'|$$

for all $x, x' \in [a, b]$.

9. Let $f: R^3 - z \to S$ be the radial projection onto the surface of a sphere from its center z (see Fig. 3.4). Let the radius of S be 1. Show that δ in the figure is given as a function of ϵ by

$$\delta = d\epsilon \sqrt{1 - \epsilon^2/4},$$

where d is the distance from x to z. Hint: Drop a perpendicular from z to the chord from fx to t, let θ denote half the angle at z determined by fx and t, and use the identity $\sin 2\theta = 2 \sin \theta \cos \theta$.

4. Open sets and closed sets

Our objective is to define and study a special class of subsets of a set X in R^n called *open* sets of X. They will play a fundamental role in our subsequent work, because the various topological properties of X we shall discuss are readily expressible in terms of the open sets. Also, the condition for a function to be continuous takes on a very simple form when open sets are used.

It is not easy to see in advance why the notion of open set should be an important concept. It is a historical fact that it gained recognition slowly. During the early development of topology (1900–1930), a variety of different approaches to the subject were devised and worked out. Attached to these are concepts with names such as: neighborhood spaces, metric spaces, limit points, sequential limits, and closures. At the time it was not clear that these approaches were equivalent; nor could one predict the direction of development and ultimate form of topology. Not until the end of this period did it gradually become clear that the concept of open set is a simple and flexible tool for the investigation of all topological properties. Since then this concept has provided the preferred approach.

DEFINITION. Let X be a subset of R^n. A subset U of X is called an *open* set of X if, for each point x of U, there is some neighborhood of x in X which lies in U. The condition may be restated: for each $x \in U$, there is a number $r > 0$ such that $N(x, r, X) \subset U$.

As we shall see shortly, open sets are easily found and occur in great variety. Our first class of examples consists of neighborhoods:

All neighborhoods are open sets.

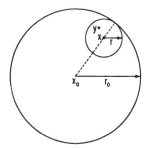

Figure 4.1

Let $x_0 \in X$ and $r_0 > 0$ be given. To prove our assertion we must show that $N(x_0, r_0, X)$ is an open set of X (see Fig. 4.1); this we shall do by showing that for each point $x \in N(x_0, r_0, X)$, there is a positive number r such that the neighborhood $N(x, r, X)$ is contained in $N(x_0, r_0, X)$. Set $r = r_0 - d(x, x_0)$. Now $x \in N(x_0, r_0, X)$ means that $d(x, x_0) < r_0$; hence r must be positive. Let $y \in N(x, r, X)$; then $d(x, y) < r$. The triangle inequality gives

$$d(x_0, y) \leq d(x_0, x) + d(x, y),$$

and since $d(x, y) < r$, we have

$$d(x_0, y) < d(x_0, x) + r.$$

But by our definition of r, $d(x_0, x) + r = r_0$, so

$$d(x_0, y) < r_0.$$

This shows that any point of $N(x, r, X)$ is a point of $N(x_0, r_0, X)$, and therefore $N(x_0, r_0, X)$ is open in X.

The following theorems show how we may manufacture additional examples of open sets from those at hand.

THEOREM 4.1. *If U and V are open sets of X, then their intersection $U \cap V$ is an open set of X. The intersection of any finite number of open sets of X is an open set of X.*

To prove the first statement, let $x \in U \cap V$. Since $x \in U$ and U is open, there is an $r > 0$ such that $N(x, r, X) \subset U$. Since $x \in V$ and V is open, there is an $s > 0$ such that $N(x, s, X) \subset V$. Let t be the smaller of r and s; it is clear that $N(x, t, X)$ lies in both U and V, and therefore in $U \cap V$. This proves that $U \cap V$ is open. To prove the second statement, suppose that x is in each of the open sets $U_1, U_2, \cdots, U_k$. Then there are numbers $r_i > 0$ ($i = 1, 2, \cdots, k$) such that $N(x, r_i, X) \subset U_i$. Take t to be the smallest of $r_1, r_2, \cdots, r_k$. Clearly $N(x, t, X)$ lies in the intersection, and this completes the proof.

In the argument given above, we assume that there is a point in $U \cap V$. If U and V have no common point, that is,

$$U \cap V = \emptyset = \text{the empty set},$$

we must test the definition of open set on the empty set. At first glance, this may seem a bit foolish; however it is strictly logical. Since $\emptyset$ has no points, it is correct to say that each of its points has a neighborhood contained in $\emptyset$. Therefore $\emptyset$ is open. Look at it the other way around. If a set A is not open in X, it contains *some* point having no neighborhood contained in A; so a non-open set is non-empty. This fact is important enough to state formally along with another important fact, namely that a set is an open set of itself.

THEOREM 4.2. *The empty set $\emptyset$ is an open set of X, and X itself is an open set of X.*

The second proposition is obvious since, for each $x \in X$,

$$N(x, r, X) \subset X$$

for all $r > 0$ by the definition of neighborhood.

The next theorem gives another method of building new open sets out of old ones.

THEOREM 4.3. *The union of any collection of open sets of X (finite or infinite in number) is an open set of X.*

To prove this, let C denote the collection of open sets, and let A denote their union. If $x \in A$, then we must have $x \in U$ for some open set U of C. Since U is open, there is an $r > 0$ such that $N(x, r, X) \subset U$. Now $U \subset A$ by the definition of the union. It follows that $N(x, r, X) \subset A$, and this shows that A is open.

These results indicate that, for most sets X, the family of open sets of X is very large. We can build open sets in endless variety by forming unions of neighborhoods. We shall now show that many familiar sets qualify as open sets.

If $X = R$, then each neighborhood is an open interval

$$N(x, r) = (x - r, x + r).$$

Each open interval is a neighborhood of its midpoint and thus is an open set of R. A union of two or more open intervals is also an open set of R; for example, the union of the sequence of non-overlapping open intervals $(1/(n+1), 1/n)$, for $n = 1, 2, 3, \cdots$, is open.

Let $X = R^2$. In this case, interiors of circles are neighborhoods so each such is open in R^2. Let A denote the set of points on four segments forming a rectangle in R^2. Then its complement $R^2 - A$ has two parts: an interior U and an exterior V (see Fig. 4.2). If x is a point of U, and we choose a positive number r less than the shortest distance from x to the sides of A, then $N(x, r)$ lies in U. Therefore U is open in R^2. Similarly, the exterior V is open in R^2. However A is not open in R^2 because it has a point z having no neighborhood $N(z, r, R^2)$ which is contained in A. In fact, every point of A has this property. These conclusions remain valid if we replace the rectangle A by any simple closed polygon such as a triangle or a hexagon.

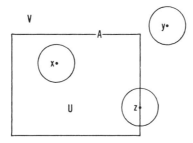

Figure 4.2

It should be noticed that the property of being open is a delicate one—it can be lost if the set is altered by a single point. In the preceding example, if we adjoin to the interior U of A a single point either from A or its exterior, then the enlarged set will not be open in R^2.

When $X = R^3$, the neighborhoods are the interiors of spheres, and each such set is open in R^3. The exteriors of spheres are likewise open in R^3, but the surface of a sphere is not open in R^3. Let A denote the set of points on the faces, edges, and vertices of a rectangular box in R^3; then $R^3 - A$ is divided into two open sets: the interior and the exterior of the box. Let T denote the set of points on the surface of a torus (doughnut) in R^3, then $R^3 - T$ is also divided into two open sets: the interior and exterior of T.

In the preceding examples, we have taken X to be all of R^n. The following theorem tells us how to "see" the open sets of a subset X once we have pictured those of R^n.

THEOREM 4.4. *If $X \subset R^n$, then the collection of open sets of X coincides with the collection of intersections of X with all open sets of R^n.*

As the first part of the proof, we shall show that, if U is an open set of R^n, then $X \cap U$ is an open set of X. Let $x \in X \cap U$. Since $x \in U$ and U is open, there is an $r > 0$ such that $N(x, r) \subset U$. Hence

$$X \cap N(x, r) \subset X \cap U.$$

But $N(x, r, X) = X \cap N(x, r)$, so

$$N(x, r, X) \subset X \cap U.$$

This says that for each $x \in X \cap U$ there is a neighborhood of x in X contained in $X \cap U$; therefore $X \cap U$ is an open set of X. This is half of what we must prove.

Next we must show that an open set V of X can be enlarged to an open set U of R^n such that $V = X \cap U$. If V were a neighborhood $N(x, r, X)$, it is clear that the desired enlargement would be $N(x, r, R^n)$. Now, it is easily verified that any open set V of X is the union of all neighborhoods contained in V, so we construct the desired enlargement of V by enlarging each neighborhood contained in V. We define U to be the union of the collection of neighborhoods $N(x, r, R^n)$ for all $x \in V$ and $r > 0$ such that $N(x, r, X) \subset V$, and we shall prove that this U fulfills the requirement, that is, we prove that U is open in R^n and that $X \cap U = V$. Since each $N(x, r, R^n)$ is open in R^n, Theorem 4.3 asserts that U is open in R^n. The proof that $X \cap U = V$ will be accomplished in two stages: first we shall show that each element of V is an element of $X \cap U$, and then that each element of $X \cap U$ is an element of V.

Since V is open in X, each point $x \in V$ has an $N(x, r, X) \subset V$, and therefore $x \in U$. This shows that $V \subset U$. But V is also a subset of X, so $V \subset X \cap U$. Finally, to show that $X \cap U \subset V$, we note that any point $y \in X \cap U$ is in both X and U. As a point of U, it lies in some $N(x, r, R^n)$ such that $N(x, r, X) \subset V$. Since it also lies in X, it is in $X \cap N(x, r, R^n) = N(x, r, X)$, and since $N(x, r, X) \subset V$, it follows that $y \in V$. This completes the proof that $X \cap U = V$ and also the proof of the theorem.

Let us summarize what has been done so far in Section 4. The main property of an open set U of X is its defining property: each $x \in U$ has a neighborhood $N(x, r, X) \subset U$. There is little more that can be said about a single open set. Our theorems state properties of the family of all open sets of X, namely, it contains as elements the empty set, X itself, and every neighborhood $N(x, r, X)$; moreover, it contains the intersection of any finite collection of its elements, and the union of any collection, finite or infinite.

We turn now to the concept of closed set.

DEFINITION. Let X be a subset of R^n. A subset A of X is called a *closed* set of X if its complement in X is an open set of X. Briefly, A is closed if $X - A$ is open.

If we refer to the definition of an open set, we obtain the following test for a set A to be closed in X: each point of $X - A$ has a neighborhood not meeting A. For example, a set consisting of a single point is always closed in any larger set X; for, if x is the point and y is any other point, then $N(y, r)$ does not contain x if r is less than or equal to the distance from x to y. Similarly, a straight line L in a plane or in space is a closed set in any larger set; for, if y is not on L, and r is the distance from y to the nearest point of L, we have that $N(y, r)$ does not meet L.

Each example we have given of an open set yields, on passing to its complement, an example of a closed set. In the example of the rectangle A of Fig. 4.2, the complement of the exterior V is the union of A with its interior U. Since V is an open set, it follows that $A \cup U$ is a closed set of R^2. Similarly, $A \cup V$ is closed in R^2. Since the union $U \cup V$ is open and A is the complement of $U \cup V$, it follows that A is closed in R^2.

The relation between a set A in X and its complement $X - A$ is reciprocal: the complement of $X - A$ is A. This correspondence between subsets of X is called *duality* in X. Open sets and closed sets are dual concepts because the dual of an open set is a closed set, and conversely.

This duality between open sets and closed sets enables us to deduce from each theorem we have proved about open sets a true "dual" theorem about closed sets. In working out the form of the dual propositions, we make use of the fact that union and intersection are "dual" operations in the following sense. *The complement of the union of two sets is the intersection of their complements:*

$$X - (A \cup B) = (X - A) \cap (X - B).$$

Similarly, *the complement of the intersection of two sets is the union of their complements:*

$$X - (A \cap B) = (X - A) \cup (X - B).$$

Thus the following four theorems for closed sets correspond to those we have proved for open sets. The theorem about the intersection of two open sets gives us the dual

THEOREM 4.1'. *If A and B are closed sets of X, then their union $A \cup B$ is a closed set of X. The union of any finite number of closed sets of X is a closed set of X.*

To obtain the dual of the proposition that $\emptyset$ and X are open, we need only observe that $\emptyset$ and X are complementary sets in X: $X - \emptyset = X$, and $X - X = \emptyset$.

THEOREM 4.2′. *Both the empty set $\emptyset$ and X itself are simultaneously open sets of X and closed sets of X.*

An open set of X is usually not a closed set of X, and vice versa. In Section 7 we shall make a careful study of subsets of X which are both open and closed in X.

The theorem on the union of any collection of open sets gives as its dual

THEOREM 4.3′. *The intersection of any collection of closed sets of X (finite or infinite in number) is a closed set of X.*

The proposition about open sets of X being the intersections of X with open sets of R^n has as its dual

THEOREM 4.4′. *If $X \subset R^n$, then the collection of closed sets of X coincides with the collection of intersections of X with all closed sets of R^n.*

Suppose $A \subset X \subset R^n$ and A is closed in R^n; then the theorem asserts that $A \cap X$ is closed in X. Since $A \cap X = A$, we obtain

COROLLARY. *If A is a closed set in R^n, then, for every set X containing A, A is a closed set in X.*

It should not be thought that every set in R^n is either open or closed in R^n; many sets are neither. The half-open interval $(a, b]$ is neither open nor closed in R. For any point x in $(a, b]$ other than b, there is some neighborhood of x which lies in the interval. On the other hand, each neighborhood of b contains points which do not lie in $(a, b]$. In the example of Fig. 4.2, the union of the interior U and a single point of A is neither an open set nor a closed set of R^2. We remarked before that it was not open. It is not closed because, for any point x of A, every $N(x, r)$ contains points of U. But by our test for a set to be closed, each point of its complement must have a neighborhood not meeting the set.

We turn now to the formulation of the continuity of a function in terms of open sets. The ease of the formulation suggests how useful open sets will be in treating questions of continuity.

THEOREM 4.5. *A function $f: X \to Y$ is continuous if and only if the inverse image of each open set of Y is an open set of X. Equivalently, f is continuous if and only if the inverse image of each closed set of Y is a closed set of X.*

Recall that f is continuous if, for each $x \in X$ and each $\epsilon > 0$, there is a $\delta > 0$ such that f maps the δ-neighborhood of x into the ϵ-neighborhood of fx. The condition that $f^{-1}V$ be open in X for each open set V of Y is surely a simpler requirement to state.

To prove the theorem, suppose first that f is continuous and that V is an open set of Y. We must show that each point $x \in f^{-1}V$ has a neighborhood contained in $f^{-1}V$. Now $fx \in V$ and V is open in Y, so there is a number $\epsilon > 0$ such that $N(fx, \epsilon, Y) \subset V$. Since f is continuous, there is a $\delta > 0$ such that the image under f of $N(x, \delta, X)$ lies in $N(fx, \epsilon, Y)$ and hence in V. Therefore $N(x, \delta, X) \subset f^{-1}V$, and this proves that $f^{-1}V$ is open for each open set V of Y.

To prove the converse, we suppose f has the property that $f^{-1}V$ is open for every open set V of Y and show that then f is continuous. Let $x \in X$ and let $\epsilon > 0$. $N(fx, \epsilon, Y)$ is an open set of Y, so its inverse image in X is open in X; denote it by U. Since x is in U and U is open, x has some neighborhood $N(x, \delta, X)$ contained in U. It follows that $fN(x, \delta, X) \subset N(fx, \epsilon, Y)$, and this shows that f is continuous.

We have proved the part of the theorem referring to open sets. The dual statement for closed sets is a consequence of the fact that for any function $f: X \to Y$ the complement in X of the inverse image of a set A of Y is the same as the inverse image of the complement of A in Y. Symbolically, for each subset $A \subset Y$,

$$X - f^{-1}A = f^{-1}(Y - A).$$

The proof of this formula is a short exercise for the reader. Let us take it for granted and suppose f is continuous and A is closed in Y. Then $Y - A$ is open in Y. By the first part of the theorem, $f^{-1}(Y - A)$ is open in X. Its complement is therefore closed in X. The formula above says that this complement is $f^{-1}A$; hence $f^{-1}A$ is closed in X.

Suppose instead that the inverse image of each closed set is a closed set. If A is an open set of Y, then $Y - A$ is closed in Y; hence $f^{-1}(Y - A)$ is closed in X. So its complement in X is open. The formula above states that this complement is $f^{-1}A$. Thus $f^{-1}A$ is open for each open set A. So f is continuous. This concludes the proof.

THEOREM 4.6. *If $f: X \to Y$ and $g: Y \to Z$ are continuous functions, then their composition $gf: X \to Z$ is continuous.*

Let W be an open set of Z. Since g is continuous, the preceding theorem asserts that $g^{-1}W$ is an open set of Y, and since f is continuous, it asserts that $f^{-1}(g^{-1}W)$ is an open set of X. It is a short exercise for the reader to verify that

$$(gf)^{-1}W = f^{-1}(g^{-1}W).$$

(See the answer to Exercise 2.8.) Thus we have shown that $(gf)^{-1}W$ is open for each open set W of Z. By the preceding theorem, this means that gf is continuous.

In subsequent sections we shall frequently use the expression "a space X". In every case, X is a subset of some R^n; however, our point of view will be that we wish to concern ourselves solely with the points of X and the open subsets of X and to ignore the surrounding space R^n. This is called the *intrinsic* viewpoint. The reader is urged to review the definitions and theorems of this section, and note that all save Theorems 4.4 and 4.4' are worded intrinsically. We shall discuss the importance of this viewpoint in Section 8

Exercises

1. If X has a finite number of points, show that every subset of X is both open and closed in X.

2. Let L be a line in R^2 and U an open interval of L; find an open set V of R^2 such that $V \cap L = U$.

3. Let D be the circular disk in R^2 of points (x, y) such that $x^2 + y^2 \leq 1$; find the largest subset of D which is open in R^2.

4. Give an example of a closed set of R^2 which becomes an open set when one of its points is deleted.

5. Give an example to show that the complement of the union of two sets is not the union of their complements.

6. Give an example to show that a union of two non-open sets can be open. (Hint: Consider half-open intervals.)

7. If $X \subset Y \subset R^n$, show that each open set of X is an intersection $X \cap V$ for some open set V of Y.

8. Let C be the collection of open intervals in R
$$I_1 = (-1, 1), \quad I_2 = (-\tfrac{1}{2}, \tfrac{1}{2}), \quad \cdots,$$
and
$$I_a = \left(-\frac{1}{a}, \frac{1}{a}\right), \quad \text{for } a = 1, 2, 3, \cdots.$$
Show that the intersection of all these open sets is not open in R.

9. Give an example of a mapping $f: X \to Y$ and a set $A \subset X$ such that $Y - fA$ is different from $f(X - A)$.

10. Prove that for any function $f: X \to Y$ the complement in X of the inverse image of a set of Y is the same as the inverse image of its complement taken in Y; that is, $X - f^{-1}A = f^{-1}(Y - A)$.

11. Show that each open set of X is the union of some collection of neighborhoods in X.

5. The completeness of the real number system

The main point of this section is that there are enough real numbers. In greater detail, if by a real number we mean something representable by a decimal expansion (finite or infinite), then there are enough real numbers to fill up the number line completely.

The history of mathematics has been marked by a succession of expansions of the number system. First there was prehistoric man with his counting: one, two, three, many. Then came the concept of the unending sequence of positive integers together with a nomenclature and an abbreviated notation. Next came the fractions or rational numbers, then came the "roots" of algebraic equations or algebraic numbers, then zero and the negative numbers, and finally the transcendental numbers.†

At each of these stages, some of those who used numbers became gradually aware of an inadequacy in the concept of numbers as they understood it. After several attempts they finally succeeded in creating new numbers which, when adjoined to the older numbers, removed the inadequacy. Most of us understand thoroughly the need for the integers and rational numbers including their negatives and zero. It is less well understood why these are not enough.

It was the school of Pythagoras which discovered that $\sqrt{2}$ is not a rational number; precisely, there is no fraction whose square is 2. Here is the proof given by Euclid. Suppose, to the contrary, that m/n is a fraction whose square is 2. We can suppose in fact that m/n is in reduced form, that is, m and n have no common integral factor other than 1. In particular then, they are not both even integers (all common 2's having been "cancelled"). We write the equation $(m/n)^2 = 2$ in the form $m^2 = 2n^2$, which tells us that m^2 is an even integer. Now the square of an odd integer is itself odd:

$$(2r + 1)^2 = 4r^2 + 4r + 1 = 2(2r^2 + 2r) + 1.$$

Since m^2 is even, it follows that m is even, so that $m = 2k$ for some integer k. If we substitute this value for m in our equation it becomes

† For a detailed treatment of the development of the number concept see Volume 1 of the New Mathematical Library, *Numbers: Rational and Irrational* by Ivan Niven.

$4k^2 = 2n^2$, hence $n^2 = 2k^2$. This means that n^2 is even, and therefore n is even. Thus both m and n are even integers, contradicting the fact that m/n was in reduced form. The contradiction shows that there can be no fraction whose square is 2.

The Pythagoreans needed the number $\sqrt{2}$ because they were geometers. Starting with a line segment of length d they could construct, using straightedge and compasses, a square of side d. By the theorem of Pythagoras, the length of the diagonal had to be $\sqrt{2}\, d$.

Think of the successive expansions of the number system this way. Starting with a line L and two points on L, called 0 and 1, one can use compasses to mark off successively the remaining integer points $2, 3, \cdots$ and $-1, -2, \cdots$. By means of another construction involving an auxiliary line, one can divide each interval $[n, n+1]$ into as many equal parts as desired. Thus all points on the line L with rational coordinates can be constructed from the two points 0 and 1.

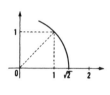

Figure 5.1　　　　Figure 5.2

Now the rational points are densely distributed along L (between any two such points there are infinitely many more distributed uniformly). It is easy to see how one might assume, without being aware of it, that these rational points were *all* the points of L. However, the Pythagoreans discovered that the diagonal of a square with side 1, when marked off on L (see Fig. 5.1), gave a point $\sqrt{2}$ which was not one of these rational points. What a jolt this must have been to the discoverers! This forced them to create new numbers to correspond to the new points of L arising from such geometric constructions.

Unfortunately, inventing the square roots of rationals still did not give enough numbers. Trisecting an angle or duplicating a cube requires the taking of cube roots of rationals, and these are not usually square roots of rationals. So mathematicians were forced into the creation of n-th roots of rationals and of the even larger set of all algebraic numbers. These are the roots of polynomial equations having integer coefficients.

About one hundred years ago it was found that the algebraic numbers are not enough; that is, there are points on the number line that do not correspond to algebraic numbers. In particular, the number π (the ratio of the circumference of a circle to its diameter) was proved to be not an algebraic number. One is led to ask: When, if ever, will the process end?

The development of the decimal system and of decimal expansions of numbers gave a new viewpoint on these questions. All of the numbers created so far can be represented by their decimal expansions. It should be emphasized that most rationals and all the other numbers have unending decimal expansions. It is natural at this stage to turn things around and to say that any decimal expansion represents a real number. That is to say, we can define the set R of real numbers to be the set of decimal expansions (with the customary convention that an expansion ending in nines represents the same number as another one ending in zeros, for example, $3.26999\cdots = 3.27000\cdots$). This is in fact what we shall do. To justify the procedure we must show that it brings an end to the game of creating new numbers; the numbers in the set R fill up the line completely.

We must explain what is meant by "filling up the line". Recall the standard method of extracting a square root such as $\sqrt{2}$. Geometrically we are dealing with the graph of the equation $y = x^2$, and we are trying to determine the x-coordinate of the point where the graph crosses the horizontal line $y = 2$ (see Fig. 5.2). We test first the squares of the first few integers and we find that $1^2 = 1$ is too small and $2^2 = 4$ is too large. Now if x is made to increase from one positive value to another, its square also increases. This fact tells us that $\sqrt{2}$ lies somewhere in the interval $I_0 = [1, 2]$, and the integer part of its decimal expansion is 1. Next we divide the interval I_0 into tenths, and square each of the numbers 1.0, 1.1, 1.2, $\cdots$, 1.9, 2.0. We find that $(1.4)^2$ is less than 2 and $(1.5)^2$ is greater than 2. Thus $\sqrt{2}$ lies in the interval $I_1 = [1.4, 1.5]$, and the decimal expansion of $\sqrt{2}$ begins with 1.4. Next we divide I_1 into ten equal parts, test the squares of the points of the division, and find that $\sqrt{2}$ lies in the interval $I_2 = [1.41, 1.42]$. Continuing thus we determine an infinite sequence of intervals $I_0 \supset I_1 \supset I_2 \supset \cdots \supset I_k \supset \cdots$ squeezing down on $\sqrt{2}$ (the symbol $\supset$ is the reverse of $\subset$, and "$A \supset B$" is read: "A contains B"). Each interval is a tenth part of the preceding, and the decimal expansion of $\sqrt{2}$ can be read off from the decimal expansions of their left-hand endpoints which are 1, then 1.4, next 1.41, etc.

Picture now a similar but more general problem; instead of $y = x^2$ consider $y = fx$, where f is any continuous function which increases in value as x increases, and instead of finding a number x such that $x^2 = 2$, we wish to solve the equation $fx = b$ where b is some given number. If we can find an initial interval $I_0 = [n, n+1]$ such that $fn < b$ and $f(n+1) > b$, then we can again carry out the procedure of repeatedly dividing intervals into tenths. This gives an infinite sequence of intervals $I_0 \supset I_1 \supset \cdots \supset I_k \supset \cdots$. Amalgamating the decimal expansions of the left-hand endpoints of these intervals just as in the case of the number $\sqrt{2}$, we can construct the decimal expansion of a number a which lies in each of the intervals, and should therefore be a solution of our problem: $fa = b$. This suggests strongly the conclusion that *we have enough real numbers to solve any problem of the above type if we agree that each decimal expansion determines a real number.*

We proceed now with the formulation and proof of a theorem which is a precise restatement of some of the preceding ideas.

DEFINITION. An infinite sequence of closed intervals of real numbers $I_0, I_1, I_2, I_3, \cdots, I_n, \cdots$ is called a *contracting* sequence if each interval contains the next and hence all intervals that follow:

$$I_0 \supset I_1 \supset I_2 \supset I_3 \supset \cdots \supset I_n \supset \cdots$$

A contracting sequence is called a *regularly* contracting sequence if $I_0 = [m, m+1]$ is the interval from an integer m to the next integer $m + 1$, and, for each $n \geq 1$, I_n is one of the intervals obtained by partitioning I_{n-1} into ten equal parts. Thus I_1 is a tenth part of I_0, I_2 a tenth part of I_1, and so on. The length of I_n is 10^{-n}.

COMPLETENESS THEOREM. *Each contracting sequence of intervals has a common point; that is, the intersection of all the intervals of the family is not empty.*

Consider first the case of a regularly contracting sequence. For each $n = 0, 1, 2, \cdots$, let a_n denote the left-hand endpoint of I_n; then $a_0 = m$ is an integer. Since I_n has length 10^{-n}, $I_n = [a_n, a_n + 10^{-n}]$. The points that divide I_{n-1} into ten equal parts are

$$a_{n-1}, \quad a_{n-1} + \frac{1}{10^n}, \quad a_{n-1} + \frac{2}{10^n}, \quad \cdots, \quad a_{n-1} + \frac{9}{10^n}, \quad a_{n-1} + \frac{10}{10^n}.$$

Since I_n is one of these ten subintervals, its left endpoint must be

$$a_n = a_{n-1} + \frac{k_n}{10^n},$$

where k_n is one of the digits $0, 1, 2, \cdots, 9$. In this way the sequence of intervals determines an integer m and an infinite sequence of digits $k_1, k_2, \cdots, k_n, \cdots$. Let c denote the real number

$$c = m + \frac{k_1}{10} + \frac{k_2}{10^2} + \cdots + \frac{k_n}{10^n} + \cdots$$

(that is, the decimal expansion of c is $m.k_1k_2k_3\cdots$). Since

$$a_n = m + \frac{k_1}{10} + \frac{k_2}{10^2} + \cdots + \frac{k_n}{10^n},$$

it is clear that $a_n \leq c$. Now the decimal expansions of $a_n + 1/10^n$ and of c agree out to the n-th digits, but the n-th digit of c is k_n while that of $a_n + 1/10^n$ is $k_n + 1$. It follows that

$$a_n \leq c \leq a_n + \frac{1}{10^n}.$$

These inequalities assert that $c \in I_n$, and since they hold for every integer n, it follows that c lies in each interval of the sequence, hence in the intersection of all of them. This proves the theorem in case the sequence contracts regularly.

Now let $I_0, I_1, \cdots$ be any contracting sequence; let $I_0 = [a_0, b_0]$, $I_1 = [a_1, b_1]$, $\cdots$, and in general, $I_n = [a_n, b_n]$. Then we have the inequalities

$$a_0 \leq a_1 \leq \cdots \leq a_n \leq \cdots \leq b_n \leq \cdots \leq b_1 \leq b_0.$$

Choose an integer $r \leq a_0$ and an integer $s > b_0$, thus obtaining an interval $I' = [r, s]$ containing I_0 and all the other intervals. The interval I' is partitioned into equal subintervals of length one by the integers between r and s. Let $I_0' = [m, m+1]$ be the subinterval such that only finitely many a's precede m, and all a's precede $m+1$. Equivalently, among those subintervals of I_0 that contain a's of the sequence, I_0' is the one farthest to the right. We now divide I_0' into ten equal parts and let $I_1' = [c_1, d_1]$ be the subinterval of I_0' such that only finitely many a's precede c_1 and all a's precede d_1. Now divide I_1' into ten equal parts and let $I_2' = [c_2, d_2]$ be the part such that only finitely many a's precede c_2, and all a's precede d_2. Continue in this manner so that, for each n, $I_n' = [c_n, d_n]$ is a tenth part of I_{n-1}', and only finitely many a's precede c_n and all a's precede d_n. Since the sequence $I_0', I_1', \cdots$ contracts regularly, there is a point c common to all the intervals. Thus $c_n \leq c \leq d_n$ for all n. Since I_n' has length 10^{-n}, we have $c_n \leq c \leq c_n + 10^{-n} = d_n$ so that

$$d_n \leq c + 10^{-n} \quad \text{and} \quad c - 10^{-n} \leq c_n.$$

We want to prove that $a_n \leq c \leq b_n$ for every n. Suppose to the contrary that, for some integer N, we have $c < a_N$. Since the powers of 10 increase without bound, we can find integers n such that

$$10^n > \frac{1}{a_N - c}.$$

Of these, choose an integer n greater than N; for such an n we have

$$a_n \geq a_N \quad \text{and} \quad 10^n > \frac{1}{a_N - c}.$$

The second inequality can be rewritten $a_N - c > 10^{-n}$ or $a_N > c + 10^{-n}$ which, combined with the first inequality, gives $a_n > c + 10^{-n}$. Since d_n does not exceed $c + 10^{-n}$, it follows that $d_n < a_n$, contradicting the fact that d_n is preceded by all the a's. The contradiction shows that

$a_n \leq c$ for all n. To prove that $c \leq b_n$ for all n, suppose to the contrary that $b_N < c$ for some integer N. Then $c - b_N > 0$, and we can choose an integer n bigger than N such that

$$10^n > \frac{1}{c - b_N}.$$

Then we have $b_n \leq b_N$ and $10^{-n} < c - b_N$. These combine to give $b_n < c - 10^{-n}$. This inequality and the one above involving c_n imply that $b_n < c_n$. Since all the a's precede all the b's, this means that all the a's precede c_n. As this is a contradiction, it follows that $c \leq b_n$ for all n. This proves that $c \in I_n$ for all n, and completes the proof of the theorem.

We could proceed now to show precisely how our completeness theorem enables us to solve equations of the type discussed earlier in this section. But all these results are embodied in the main theorem of Part I (see Section 1); so we shall continue with the working out of its proof.

Exercises

1. Show that $\sqrt{3}$ is not a rational number. [Hint: Show that an integer whose square is divisible by 3 is also divisible by 3; or equivalently, show that the square of an integer not divisible by 3 (i.e., of the form $3k + 1$ or $3k + 2$) is not divisible by 3.]

2. Give another proof of the proposition that the equation $2n^2 = m^2$ has no solution in integers m, n, using the theorem that each integer has a unique factorization as a product of prime numbers. (Hint: Compare the number of factors 2 on each side of the equation.)

3. Show that $\sqrt[3]{2}$ is not a rational number by proving (with the aid of the unique prime factorization theorem) that $2n^3 = m^3$ has no solution in integers.

4. Show that the equation $y^2 = 2x^2$ has no solution in rational numbers x, y other than $(0, 0)$.

5. Show that there are no integers k, m and n such that

$$[(k + m\sqrt{2})/n]^3 = 2.$$

6. If $S = a_1 + a_2 + a_3 + \cdots$ is an infinite series, the finite sums

$$S_0 = 0, \quad S_1 = a_1, \quad S_2 = a_1 + a_2, \quad \cdots,$$
$$S_n = a_1 + a_2 + \cdots + a_n, \quad \cdots$$

are called *partial* sums of the series. Show how the partial sums of the infinite series

$$1 - \frac{1}{10} + \frac{1}{10^2} - \frac{1}{10^3} + \frac{1}{10^4} - \cdots + (-1)^n \frac{1}{10^n} + \cdots$$

determine a regularly contracting sequence of intervals. What is the sum of the series?

7. Show how the partial sums of the infinite series

$$1 - \frac{1}{2} + \frac{1}{2^2} - \frac{1}{2^3} + \cdots + (-1)^n \frac{1}{2^n} + \cdots$$

determine a contracting sequence of intervals whose intersection is the sum of the series. What is this sum?

8. Show how the process of long division applied to 12.27/3.41 leads to a regularly contracting sequence of intervals.

9. Find $\sqrt[3]{4}$ to the second decimal place by the method of contracting intervals.

10. Show that each open interval (a, b), where $a < b$, contains a rational number and also an irrational number; show that the set Q of all rational numbers is neither closed nor open in R.

11. Prove the theorem of Dedekind about a "cut" of the real numbers: If A and B are two non-empty subsets of R such that $R = A \cup B$ and every number of A is less than every number of B, then there is a real number which is either the largest number of A or the smallest number of B.

6. Compactness

A subset X of R^m is said to be *bounded* if it is contained in some sufficiently large ball; that is, if there is a point x_0 and a number $r > 0$ such that $X \subset N(x_0, r)$. Examples of bounded sets are segments, circles, spheres, triangles, etc. Examples of unbounded sets are lines, half-lines, rays, planes, exteriors of circles in R^2, the entire space R^m, and the rational numbers. Intuitively, a set is unbounded if one can run off to infinity along the set.

A most important and remarkable property possessed by any subset X of R^m that is both closed in R^m and bounded is that, for any continuous mapping $f \colon X \to R^m$, the image set fX is also closed and bounded. The main purpose of this section is to prove this fact. The proof is necessarily somewhat indirect; its development passed through many stages beginning with the work of Cauchy (1789–1857). In particular, it embodies the propositions of analysis often referred to as the theorem of Bolzano–Weierstrass and the theorem of Heine–Borel.

In the proof to be given, we first show that the property of being closed and bounded is equivalent to another property called "compactness". This is the major part of the argument. Once this is done, it is easy to show that fX is compact whenever X is compact and f is continuous. We shall lead up to the definition of compactness by displaying a common property of unbounded sets and non-closed sets.

Let X be an unbounded subset of R^m and x_0 a point of R^m. Picture the sequence of neighborhoods $N(x_0, r)$ where the radius r takes on the values $r = 1, 2, 3, \cdots$. These form an expanding sequence of open sets whose union is all of R^m, since, for each $x \in R^m$, the distance $d(x, x_0)$ is less than r for some sufficiently large integer r. The intersections $X \cap N(x_0, r)$, $r = 1, 2, 3, \cdots$, form therefore an expanding sequence of open sets of X whose union is all of X; but X is not equal to any one of these open sets because X is unbounded. Moreover, X is not contained in the union of any finite number of these sets because their union is just the largest one.

Now let X be a bounded but non-closed subset of R^m; then there is some point y in the complement $R^m - X$ of X such that each neighborhood $N(y, r)$ contains points of X (see the definition of closed set in Section 4). For each integer $k = 1, 2, 3, \cdots$, let U_k be the exterior of the circle about y of radius $1/k$. Each U_k is an open set of R^m; for, if $x \in U_k$, then $N(x, d(x, y) - 1/k)$ is a neighborhood of x contained in U_k. The U_k form an expanding sequence $U_1 \subset U_2 \subset \cdots$, and their union is the complement of y, because for each point $x \neq y$ there is a k such that $(1/k) < d(x, y)$. It follows that the intersections $X \cap U_k$ form an expanding sequence of open sets of X whose union is all of X, but X is not equal to any one of the sets because each $N(y, 1/k)$ contains points of X. Moreover, X is not in the union of any finite number of the sets because their union is just the largest one.

Thus if X is unbounded or if X is not closed we can find in X an expanding sequence of open sets of X whose union is X, but X is not the union of any finite number of them. This leads us to the definition of compactness. First, however, we need the definition of an "open covering".

DEFINITIONS. Let X be a subset of R^m. A collection C of subsets of R^m is called a *covering* of X if the union of the sets of C contains X; that is, each point of X lies in at least one of the sets of C. A covering C of X is called *finite* if the number of sets in C is finite. A covering C of X is said to *contain* a covering D of X if each set of D is also a set of C. A covering of X is called an *open* covering of X if each set of the covering is an open set of X. Finally, the space X is called *compact* if each open covering of X contains a finite covering of X; that is to say, from any infinite collection of open sets of X, whose union is X, we can select a finite subcollection whose union is also X.

If X is unbounded or not closed in R^m, the expanding sequence of open sets constructed above is an open covering of X. Any finite number of them has as their union the largest of them. Since no one of them is all of X, it follows that X is not covered by any finite number of them; hence X is not compact. We now state this result in its positive form:

THEOREM 6.1. *Every compact subset of R^m is bounded and closed in R^m.*

Eventually we must prove also the converse: every closed and bounded subset of R^m is compact. This is more difficult and will be accomplished in stages. (At the moment, the only sets that are obviously compact are those with only a finite number of points: choose one set of the covering containing each point.) The first stage and the first non-trivial case is that of an interval:

Any closed interval $I = [a, b]$ of real numbers is compact.

To prove this, we assume, to the contrary, that C is an open covering of I which contains no finite covering of I, and deduce a contradiction. Under this assumption we shall construct a contracting sequence of intervals $I = I_0 \supset I_1 \supset I_2 \supset \cdots$ such that each interval is a half of its predecessor, and no one of them is covered by a finite subcollection of C. By means of the completeness theorem of Section 5, we shall then show that one of the intervals I_k is, in fact, contained in a single set of C, a contradiction.

To construct the contracting sequence, we note that $I_0 = I$ is not finitely covered by C by hypothesis. The midpoint of I_0 divides I_0 into two closed intervals I_0' and I_0'' whose union is I_0. At least one of I_0', I_0'' is not finitely covered by C; for if C contained finite coverings C' of I_0' and C'' of I_0'', then $C' \cup C''$ would be a finite covering of I_0. Choose the half of I_0 not finitely covered by C and call it I_1. (In case both are not finitely covered, choose the right-hand half so as to make the selection specific.) We now divide I_1 in half and proceed as before. If $I_0, I_1, \cdots, I_{k-1}$ have been properly constructed, we argue as above that, since I_{k-1} is not finitely covered by C, at least one of its halves is not finitely covered by C. We select such a half and call it I_k. This completes the inductive proof of the existence of the contracting sequence.

By the completeness of R (see Section 5), there is a point x such that $x \in I_k$ for all k. Since $x \in I$ and C covers I, there is an open set U of the open covering C such that $x \in U$. Hence there is a number $r > 0$ such that $N(x, r, I) \subset U$. Now the intervals

$$I_0, I_1, \cdots, I_k, \cdots$$

contain x and have decreasing lengths

$$(b - a), \quad (b - a)/2, \quad \cdots, \quad (b - a)/2^k, \quad \cdots.$$

If we choose a k big enough so that $(b - a)/2^k < r$, then I_k will lie entirely in $N(x, r, I)$. Here is our contradiction: there is a k such that

$$I_k \subset N(x, r, I) \subset U,$$

so I_k is covered by a single set of the collection C, and yet each interval of our sequence is not covered by any finite subcollection of C. This contradiction shows that I is compact.

Before starting on our next case, we need another definition. A subset B of R^m is called an *m-dimensional box* if there are pairs of numbers $a_i < b_i$ for $i = 1, 2, \cdots, m$ such that B consists of all points x whose coordinates $(x_1, \cdots, x_m)$ satisfy the conditions $a_i \leq x_i \leq b_i$ for $i = 1, \cdots, m$. In case $m = 1$, B is just an interval. When $m = 2$, B is a rectangle and its interior, with sides parallel to the coordinate axes. When $m = 3$, B is a rectangular box and its interior, with its faces parallel to the coordinate planes.

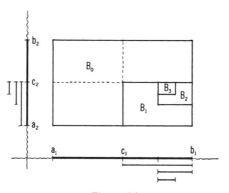

Figure 6.1

We need also to be able to *subdivide* the box B into smaller boxes. This is done by dividing each of the intervals $[a_i, b_i]$, $i = 1, 2, \cdots, m$, into two intervals by its midpoint c_i; each box of the subdivision has as its i-th interval either $[a_i, c_i]$ or $[c_i, b_i]$. Thus when $m = 2$, the rectangle is divided by the lines $x_1 = c_1$ and $x_2 = c_2$ into $4 = 2^2$ congruent rectangles, with edges half as long as those of B (see Fig. 6.1). When $m = 3$, the box is divided by the three planes $x_1 = c_1$, $x_2 = c_2$ and $x_3 = c_3$ into $8 = 2^3$ congruent boxes, with edges half as long as those of B. In general, B is divided by the m hyperplanes

§6] COMPACTNESS 41

$x_i = c_i$ ($i = 1, 2, \cdots, m$) into 2^m congruent boxes, with edges half as long as those of B.

Any m-dimensional box B is compact.

We shall prove this by a process similar to that used for the one-dimensional box, the interval. Assume, to the contrary, that C is an open covering of B which contains no finite covering. We shall construct a contracting sequence of boxes $B_0, B_1, \cdots, B_k, \cdots$ such that $B_0 = B$, no one of them is covered by a finite subcollection of C, and, for each $k > 0$, B_k is one of the 2^m boxes of the subdivision of B_{k-1}. By hypothesis $B_0 = B$ is not finitely covered by C. This starts our inductive construction of the sequence. Assuming that $B_0, B_1, \cdots, B_{k-1}$ have been properly chosen, consider the 2^m boxes of the subdivision of B_{k-1}. If each were covered by a finite subcollection of C, then the union of these 2^m subcollections would be a finite subcollection of C covering B_{k-1}. Since this is impossible, at least one of these subboxes of B_{k-1} is not finitely covered by C. Choose B_k to be one such box. This completes the inductive proof of the existence of the sequence $B_0, B_1, \cdots, B_k, \cdots$. (Fig. 6.1 illustrates the first three stages for $m = 2$.)

We claim that there is a point $x \in B_k$ for every $k = 1, 2, \cdots$. To see this, consider the projections of the sequence of boxes on the i-th coordinate axis, for $i = 1, 2, \cdots, m$. On each axis the projections form a contracting sequence of intervals. Let x_i be a number common to all the intervals formed by the projections on the i-th coordinate axis. Then the point x of R^m whose coordinates are $(x_1, x_2, \cdots, x_m)$ is a point of B_k for all k's. Since $x \in B$, there is an open set U of the covering C such that $x \in U$. Hence there is a number $r > 0$ such that $N(x, r, B) \subset U$. Let d denote the length of the longest edge of B. Since all edges were bisected at each stage of the construction of the sequence, it follows that $d/2^k$ is the length of the longest edge of B_k. By the theorem of Pythagoras, the length of the diagonal of B_k is at most $\sqrt{m}\, d/2^k$. Choose an integer k so large that

$$2^k > \frac{\sqrt{m}\, d}{r}; \quad \text{then} \quad \frac{\sqrt{m}\, d}{2^k} < r,$$

and it follows that

$$B_k \subset N(x, r, B) \subset U;$$

hence B_k is contained in a single set of C, and this contradicts the fact that B_k is not finitely covered by C. Our supposition that B is not compact has led to a contradiction; therefore B is compact.

The collection of sets which we can prove to be compact is greatly enlarged by the following useful proposition.

THEOREM 6.2. *If X is a closed subset of a compact space B, then X is compact.*

To prove this, we shall take any open covering C of X and enlarge each member of C so that the enlarged sets form an open covering C' of B. We then use the compactness of B to select a finite covering from C', and observe that the corresponding unenlarged sets form the desired finite covering of X.

For each set U of the open covering C of X, let $U' = U \cup (B - X)$, and let C' denote the collection of these larger sets U'. First we shall show that U' is an open set of B. A point $x \in U'$ is either in U or in $B - X$. If $x \in B - X$, the hypothesis that X is closed in B tells us that there is an $r > 0$ such that $N(x, r, B) \subset B - X \subset U'$. If $x \in U$, the fact that U is open in X means that there is an $r > 0$ such that $N(x, r, X) \subset U$, and hence $N(x, r, B) \subset U \cup (B - X) = U'$. This proves that U' is open in B.

Now let y be any point of B; either $y \in X$ or $y \in B - X$. If $y \in X$, then $y \in U$ for some $U \in C$ so that $y \in U'$ for the corresponding $U' \in C'$. If $y \in B - X$, then $y \in U'$ for all $U' \in C'$. Therefore C' is an open covering of B. Since B is compact, a finite number of the sets of C', say $U_1', U_2', \cdots, U_k'$ cover B and hence also X. It follows that the corresponding sets of C, namely, $U_1, U_2, \cdots, U_k$, form a finite covering of X; for every point of X covered by U_j' is also covered by U_j. This completes the proof of the theorem.

We are now able to prove the converse of Theorem 6.1 by enclosing our closed and bounded set in an m-dimensional box (which has been shown to be compact) and then applying Theorem 6.2.

THEOREM 6.3. *Each closed and bounded subset of R^m is compact.*

Let X be closed in R^m and bounded. Since X is bounded, there is a point $b \in R^m$ and a number $r > 0$ such that $X \subset N(b, r)$. Let B denote the m-dimensional box with center at b and with edges all equal to $2r$; precisely, a point $y \in R^m$ is in B if its coordinates $(y_1, \cdots, y_m)$ satisfy

$$b_i - r \leq y_i \leq b_i + r \quad \text{for } i = 1, \cdots, m.$$

Then B contains $N(b, r)$, and therefore $B \supset X$; hence $X \cap B = X$. Furthermore X is closed in B; for X is closed in R^m, and by Theorem 4.4', the closed sets of B coincide with the intersections of B with the closed sets of R^m. The desired conclusion, that X is compact, is now a consequence of the preceding theorem.

At this point we have established the equivalence of the property of being compact and the property of being closed and bounded in R^m. We are prepared now to prove the main proposition of this section.

THEOREM 6.4. *Let X be a compact space and let $f: X \to Y$ be continuous; then the image fX is compact.*

PROOF. Let C be an open covering of fX. We want to show that C contains a finite covering of fX. For each $U \in C$, consider the inverse image $f^{-1}U$, and let C' be the collection of all such inverse images. Since f is continuous and U is an open set of fX, each $f^{-1}U$ is an open set of X. For each $x \in X$, fx lies in some $U \in C$ because C covers fX; so x lies in the corresponding $f^{-1}U$. Thus C' is an open covering of X. Since X is compact, there is a finite subcollection D' of C' which covers X. The corresponding subcollection D of C is finite and covers fX; for, if $x \in f^{-1}U$ and $f^{-1}U$ is in D', then $fx \in U$ where U is in D. Thus C contains a finite covering of fX. This completes the proof that fX is compact.

An immediate consequence of Theorem 6.4 is the following.

COROLLARY. *If X is a closed and bounded set in R^m, and if $f: X \to R^n$ is continuous, then fX is a closed and bounded set in R^n.*

To relate the preceding work to the main objective of Part I (to prove the theorem of Section 1), we must prove an important property of compact sets on a line.

THEOREM 6.5. *A compact non-empty set X of real numbers has a maximum and a minimum; that is, there are numbers m and M in X such that m is the smallest number in X and M is the largest number in X.*

To appreciate the force of the conclusion, note first that the set R of all real numbers has no largest number and no smallest. In fact, any unbounded set Y of numbers must fail to have a maximum or a minimum; for, if it had both, any open interval containing both would contain all of Y, and then Y would be bounded. Note next that there are bounded sets which have neither a maximum nor a minimum. For example, an open interval (a, b) has neither a largest number nor a smallest number. These examples show that, to achieve the conclusion of the theorem, we must require X to be bounded, and we must impose some additional condition which is not satisfied by an open interval. Since a compact set is bounded and closed, the single condition of compactness guarantees boundedness and rules out the open interval.

Let us proceed with the proof of the theorem. Since X is compact, it is bounded; hence there is a closed interval $I_0 = [a_0, b_0]$ which contains X. We shall construct a contracting sequence of intervals $I_0, I_1, \cdots, I_k, \cdots$ with the following properties: each interval I_k is a half of I_{k-1}; each I_k contains at least one point of X; and finally, the right-hand endpoint b_k of I_k is an upper bound of X—that is, for all $x \in X$, we have $x \le b_k$, $k = 0, 1, 2, \cdots$. Clearly I_0 contains a point of X (since X is not empty), and b_0 is an upper bound of X. Assume that $I_0, I_1, \cdots, I_{k-1}$ have been properly selected. Let c be the midpoint of $I_{k-1} = [a_{k-1}, b_{k-1}]$. If c is an upper bound of X, we take $I_k = [a_{k-1}, c]$, and if c is not an upper bound, we set $I_k = [c, b_{k-1}]$. In either case, I_k has the required properties.

By the completeness of R (see Section 5), there is a number M such that $M \in I_k$ for every $k = 0, 1, 2, \cdots$. We shall show first that M belongs to X. Suppose the contrary were true. Since X is a closed set, the complement $R - X$ is open; then there would be an $r > 0$ such that $N(M, r) \subset R - X$. If d denotes the length of I_0, then $d/2^k$ is the length of I_k. For a sufficiently large integer k we have $d/2^k < r$, so I_k contains M and has length less than r; hence

$$I_k \subset N(M, r) \subset R - X.$$

This contradicts the fact that each I_k contains a point of X. It follows that M must belong to X.

Now we show that M is the largest number of X. Suppose to the contrary that there were an $x \in X$ such that $x > M$. Take $r = x - M$ so that $r > 0$, and choose an integer k so large that $d/2^k < r$. Since $M \in I_k$ and the length of I_k is less than r, it follows that $b_k < x$; so b_k is not an upper bound of X. But b_k is an upper bound by construction. This contradiction shows that M is the largest number of X.

The proof of the existence of the minimum proceeds similarly. The contracting sequence of intervals is chosen so that each interval contains some point of X, and the *left*-hand endpoint of each is a *lower* bound of X. The details of the proof are left to the reader. One can also obtain the existence of the minimum from that of the maximum by applying the mapping $f: R \to R$ defined by $fx = -x$. Since X is compact, Theorem 6.4 asserts that fX is compact. Then fX has a maximum, say M'. It follows that fM' is the required minimum of X.

Our final theorem of this section provides a part of the conclusion of the main theorem stated in Section 1.

THEOREM 6.6. *If X is a closed, bounded and non-empty subset of R^m, and if $f: X \to R$ is a continuous real-valued function defined on X, then the image fX has a maximum M and a minimum m.*

Since X is closed and bounded, it is compact. Since X is compact and f is continuous, the image fX is compact. Since fX is a compact and non-empty set of real numbers, the preceding theorem assures us of the existence of m and M. This completes the proof.

Exercises

1. Show that any subset of a bounded set is bounded.

2. Show that the union of two bounded sets is bounded, also that the union of a finite number of bounded sets is bounded.

3. Find an example of an infinite sequence of bounded sets whose union is unbounded.

4. Find an expanding sequence of open subsets $U_1, U_2, \cdots, U_k, \cdots$ of the half-open interval $X = [a, b)$ whose union is X but no one of them is all of X.

5. Let D be the disk in R^2 with center x_0 and radius 1; that is, D consists of all points $x \in R^2$ such that $d(x, x_0) \leq 1$. If X is the set obtained by deleting x_0 from D, solve the preceding problem for this X. Do the same if X is obtained by deleting from D a point y_0 of its boundary, i.e., $d(y_0, x_0) = 1$.

6. Let X be the closed interval $[0, 10] \subset R$. Show that the set C of all open intervals of R of length 1 is a covering of X. Find a finite subcollection of C covering X. What is the least number of such intervals in a covering of X?

7. Let C_r denote the circle in R^2 with center at x_0 and radius r. For each point c on the unit circle, that is $c \in C_1$, let T_c be the line through c tangent to C_1. Then $R^2 - T_c$ is divided into two open half-planes; we let U_c be the one not containing x_0. Show that the collection C of these half-planes U_c for all $c \in C_1$ covers the exterior of C_1. If $r > 1$, why must there be a finite subcollection of C covering C_r? For $r > 2$, show that C_r can be covered by three sets of C but not by two. For r such that $\sqrt{2} < r \leq 2$, show that C_r can be covered by 4 sets of C but not by 3. For r such that $2/\sqrt{3} < r \leq \sqrt{2}$, show that C_r can be covered by 6 sets of C but not by 4. Can the open annulus between C_1 and C_2 be covered by a finite subcollection of C? (An annulus is the ring between two concentric circles.)

8. Show that the union of two compact sets is a compact set. Similarly, the union of any finite number of compact sets is compact.

9. Give an example of an infinite collection of compact sets whose union is not compact.

10. If X is compact and $X \subset Y$, show that X is closed in Y.

11. Find a bounded subset X of the rational numbers Q such that X is closed in Q, but X contains neither a maximum nor a minimum.

12. Show that there is a continuous mapping of the interval $I = [-1, 1]$ onto the interval $[-n, n]$ for each positive integer n. Does there exist a continuous mapping of I onto the entire real line R? Construct a continuous mapping of the open interval $(-1, 1)$ onto the entire line R.

13. Show that a subset X of R^m is compact if and only if each covering of X by open sets of R^m contains a finite covering.

7. Connectedness

For the proof of our main theorem of Part I, we need two important topological properties of the closed interval. The first of these, compactness, has been treated in Section 6. We shall discuss now the other property called "connectedness".

Some spaces can be divided in a natural way into two or more parts. For example, a space consisting of two non-intersecting lines can be divided into the two lines. As another example, the complement of a circle in the plane consists of two parts, the part inside the circle and the part outside. Again, if p is a point of a line L, then the complement of p in L falls naturally into the two half-lines determined by p (the deletion of p cuts L into two parts).

In each of the foregoing examples, the natural division occurs in just one way. The set Q of rational numbers can be divided into parts in many ways. Each irrational number x produces a division of Q into those rationals greater than x and those less than x. The set of irrational numbers can be divided by each rational number in a similar manner.

On the other hand, certain sets cannot be divided into parts in any natural way; this is true, for example, of a line, a line segment, a plane, and a circular disk.

Of course it is possible to force a division. For example, if I is the interval $[a, b]$ and if c is a number such that $a < c < b$, then c divides I into the two intervals $[a, c]$ and $[c, b]$. However, since they have c in common, we do not regard this as a proper division. We obtain a proper division by deleting c from one of the sets, say the second. Let $A = [a, c]$ and $B = (c, b]$; then $A \cup B = I$ and $A \cap B = \emptyset$. We do not regard such a division or "breaking" of I as natural because the set B "sticks" to A at the point c. If we delete c also from A to overcome this "stickiness", then $A \cup B \neq I$, but $A \cup B$ is the complement of c in I; hence this example is similar to the example of the complement of a point p in a line L.

The precise notion we need is now stated.

DEFINITION. A *separation* of a space X is a pair A, B of non-empty subsets of X such that $A \cup B = X$, $A \cap B = \emptyset$, and both A and B are open in X. A space which has no separation is said to be *connected*.

Consider, for example, the complement X of a circle C in the plane. Let A be its *interior* and B its *exterior*; that is, A consists of all points of X whose distance from the center of C is less than the radius of C, and B is the complement of A in X. The conditions for a separation are readily verified. The fact that A and B are open in X is apparent from Fig. 7.1; each point of A has a neighborhood contained in A, and each point of B has a neighborhood in B.

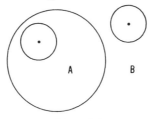

Figure 7.1

Let L be a line, p a point of L, and X the complement of p in L. Let A be the set of points of L to the left of p (see Fig. 7.2) and B the set of points to the right of p. Again each point of A has a neighborhood in A, and similarly for B.

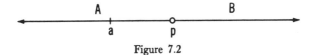

Figure 7.2

Recall now the "forced" division of the interval $I = [a, b]$ into $A = [a, c]$ and $B = (c, b]$. If we check the conditions for a separation we find that all except one are satisfied: A is not open in I because no neighborhood of the point $c \in A$ lies entirely in A.

These preliminary considerations indicate that the definitions of a separation and of a connected space express precisely the rough geometric idea we have in mind, and the next theorems will justify these definitions completely.

The definition of a separation can be reworded in several equivalent ways. Since A and B are complements of each other in X, each is open if and only if the other is closed. Thus we could equally well require that A and B be closed in X. Also we may drop explicit reference to B, and say that a separation of X is a subset A of X which is both an open and a closed set of X, and which is neither $\emptyset$ nor X. Then its complement B in X has the same properties. (Recall that $\emptyset$ and X are both open and closed in X.)

Thus, any of the following may serve as definition of a separation A, B of a space X:

1. A and B are non-empty subsets of X such that $A \cup B = X$, $A \cap B = \emptyset$, and both A and B are open in X;

2. A and B are non-empty subsets of X such that $A \cup B = X$, $A \cap B = \emptyset$, and both A and B are closed in X;

3. A is a subset of X which is both an open and a closed set of X, and which is neither $\emptyset$ nor X.

It is usually easier to prove that a space is not connected than to prove that it is connected. In the first case, we need only exhibit a separation and verify that it is one, while, in the second case, we must prove something about *all* open sets of X other than $\emptyset$ and X, namely, that each such set is not closed in X. The following theorems not only show that certain simple spaces are connected, but also present a technique for verifying the connectedness of many spaces.

THEOREM 7.1. *A closed interval of real numbers is a connected set.*

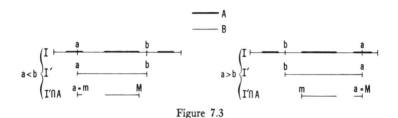

Figure 7.3

Let I be a closed interval of R, and let A be a closed set of I which is neither $\emptyset$ nor I. To prove the theorem we shall show that A is not open in I. Since $A \neq \emptyset$ and $A \neq I$, there is a point $a \in A$ and a point $b \in I - A$. Let I' denote the interval $[a, b]$ (or $[b, a]$, if $b < a$). To picture this, imagine the interval I made up of the subsets A and B shown in Fig. 7.3 where A is a closed set and B is its complement in I. Since A and I' are closed, so is their intersection $A \cap I'$. Since $A \cap I'$ is also bounded, it is a compact set. By Theorem 6.5, $A \cap I'$ has a minimum m and a maximum M. If $a < b$, then $m = a$ since a is the left-hand endpoint of I'. Since the right-hand endpoint b is not in A, we have $m = a \leq M < b$. It follows that each neighborhood of M contains numbers between M and b; these are not in A, hence A is not open. If $a > b$, then $M = a$, $b < m \leq a$, and each neighborhood of m contains numbers between b and m; these are not in A, so again A is not open. This proves that the only subsets of I which are both open and closed are I and $\emptyset$. Therefore I is connected.

THEOREM 7.2. *If $f: X \to Y$ is a continuous map and A, B is a separation of the image fX, then the inverse images $A' = f^{-1}A$ and $B' = f^{-1}B$ form a separation of X.*

We must verify that the pair A', B' satisfies each of the conditions for a separation. Since A is not empty, there is a point $y \in A$. Since

$A \subset fX$, there is an $x \in X$ such that $fx = y$; so $x \in A'$, and hence A' is not empty. Similarly, B' is not empty. Any point $x \in X$ has its image in $fX = A \cup B$, so $fx \in A$ or $fx \in B$. Accordingly, $x \in A'$ or $x \in B'$. This shows that $X = A' \cup B'$. If there were a point x common to A' and B', it would follow that fx is common to A and B. This is impossible, therefore $A' \cap B' = \emptyset$. Finally, since A and B are open in fX and f is continuous, their inverse images are open in X (see Theorem 4.5). This proves that A', B' is a separation of X.

The theorem can be restated briefly: *If f is continuous and fX is not connected, then X is not connected*. Hence an equivalent assertion is:

COROLLARY. *If $f: X \to Y$ is continuous and X is connected, then fX is connected.*

This is the form of the proposition most useful to us; for example, it enables us to prove:

COROLLARY. *Each line segment is a connected set.*

PROOF. If L is a line segment and I is an interval, there is a similarity $f: I \to L$ so that $fI = L$. Since a similarity is continuous (see Section 3) and I is connected, it follows that L is connected.

It is intuitively evident that two line segments meeting at a point together form a connected set. The proof of this rests on the following two lemmas.

LEMMA 7.3. *If X is not connected and A, B is a separation of X, then each connected subset of X lies wholly in A or wholly in B.*

Suppose C is a subset of X which contains a point of A and a point of B. Then $C \cap A$ and $C \cap B$ are not empty, each is an open set of C, their union is C, and their intersection is empty. Therefore C is not connected. This shows that a connected subset could not contain points of both A and B.

LEMMA 7.4. *If two connected sets Z and W have a point in common, then $Z \cup W$ is connected.*

Suppose, to the contrary, that $Z \cup W$ has a separation A, B. Let c be a point common to Z and W. In case $c \in A$, then both Z and W are connected sets containing a point of A. By Lemma 7.3 (with $X = Z \cup W$), Z and W lie wholly in A. So B must be empty. In case $c \in B$, we find that Z and W lie wholly in B, and then $A = \emptyset$. In either case we have a contradiction; therefore $Z \cup W$ is connected.

A simple application of Lemma 7.4 shows that two line segments meeting at a point together form a connected set; by attaching additional segments one at a time, we can conclude that any polygonal path is a connected set.

The next theorem is an important tool for showing that certain spaces are connected.

THEOREM 7.5. *A space X is connected if and only if each pair of points of X lies in some connected subset of X.*

The proof of the part of the theorem stating that, if X is connected, then each pair of points of X lies in some connected subset of X is a triviality, for the whole set X is a connected subset of itself containing every pair of its points.

To prove the other half of the theorem, assume that X has the property that each pair of its points lies in a connected subset of X, and suppose that X is not connected. Let A, B be a separation of X, and let $x \in A$ and $y \in B$ (recall that A and B are not empty). By hypothesis there is a connected set C in X containing x and y. However, by Lemma 7.3, C lies wholly in A or wholly in B. This contradiction shows that X can have no separation. Therefore X is connected, and the theorem is proved.

Recall that a subset of R^n is called *convex* if it contains all line segments joining all pairs of its points (for example, the interior of a circle in the plane is convex, but its exterior is not). Since line segments are connected, Theorem 7.5 implies

COROLLARY. *Each convex set is connected.*

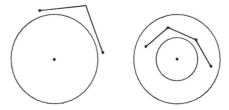

Figure 7.4

A set need not be convex to be connected. Although the exterior of a circle is not convex, it is nevertheless connected because any two of its points may be joined by a polygonal path lying in the exterior (see Fig. 7.4). Similarly, any two points of an annular region between two circles may be connected by a polygonal path; hence an annulus is connected.

We return now to the study of subsets of a line (that is, subsets of the real numbers) for our final proposition about connectedness.

§7] CONNECTEDNESS

THEOREM 7.6. *A compact connected set of real numbers is a closed interval.*

The converse of this theorem, that a closed interval is both compact and connected, has already been proved (see Section 6 and Theorem 7.1).

PROOF. Let X be any connected set of real numbers and let a and b be numbers in X with $a < b$. We prove first that any number c such that $a < c < b$ is also in X. The point c determines a separation of its complement in R; let A consist of all numbers less than c, and let B consist of all numbers greater than c. If, contrary to our claim, X did not contain c, it would be a subset of $A \cup B$, and, since X is connected, by Lemma 7.3 it would lie wholly in A or wholly in B. But X contains a and b, so this is impossible. Thus we have shown that *a connected set of real numbers contains all numbers between any two of its numbers.*

If, in addition, X is compact, Theorem 6.5 asserts that X has a minimum m and a maximum M. It follows that X is precisely the closed interval $[m, M]$.

Exercises

1. State whether each of the following sets is connected; if not connected, find a separation.

   (a) A circle with one point deleted; with two points deleted.

   (b) An arc of a circle; an arc with its midpoint deleted.

   (c) A finite set of points; the singleton set consisting of a single point; the empty set.

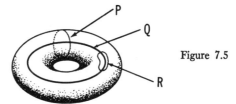

Figure 7.5

   (d) The torus (see Fig. 7.5)
      (i) with circle P deleted;
      (ii) with circle Q deleted;
      (iii) with circles P and Q deleted;
      (iv) with closed curve R deleted;
      (v) with two circles of type P deleted;
      (vi) with two circles of type Q deleted;
      (vii) with its interior, but with two circles of type P deleted.

(e) The union of two disjoint circles in the plane; the intersection of these two disjoint circles.

(f) Let A, B, C, D be four points on a circle, equally spaced and in order. Letting AB denote the shortest arc from A to B including the end points, answer the question for the following sets:

(i) $AB \cup BC$;   (ii) $AB \cap BC$;   (iii) $AB \cup CD$;

(iv) $AB \cap CD$;   (v) $ABC \cup CDA$.

2. A subset D of R^m is called *star-shaped* about a point p if, for each point $x \in D$, the line segment p to x lies in D. Show that such a set is connected.

3. Show that each of the following is connected: the surface of a sphere; the interior of a sphere; the exterior of a sphere; the surface of a torus; the interior of a torus; the exterior of a torus.

4. Show by an example that the inverse image of a connected set is not necessarily connected.

5. Show that central projection of a non-diametral chord or of a tangential line segment into a circle is continuous; conclude that circular arcs are connected.

6. Give an example of two connected sets whose intersection is not connected.

7. Explain whether the points in the plane having at least one rational coordinate form a connected set; those having exactly one rational coordinate; those having two rational coordinates. If not connected, show a separation.

8. Let X be the set of points on all circles in the plane with center at $(0, 0)$ and with radius r, where r is a rational number; find a separation of X.

9. Show that a connected set of real numbers is one of eight things: the empty set, R itself, an open or closed half-line, a single point, or an open, closed, or half-closed interval.

10. Show that the intersection of a contracting sequence of closed intervals is either a single point or a closed interval.

11. Give another proof that any interval I is connected by assuming that there is a separation $I = A \cup B$, constructing a contracting sequence of intervals each with one end in A and the other in B, and deducing a contradiction by showing that a point c of their intersection is not in A or in B.

8. Topological properties and topological equivalences

The principal task in these sections is the proof of the main theorem stated in Section 1: If the real-valued function fx is defined and continuous for $a \leq x \leq b$, then it has a minimum value, a maximum value, and takes on all values between. We have now completed all the work required for the proof; it is only necessary to assemble its various parts. The following three propositions have been proved:

1. *A closed interval of real numbers is a compact and connected set.*

2. *A continuous image of a compact set is compact, and a continuous image of a connected set is connected.*

3. *A compact and connected set of real numbers is a closed interval.*

Each of the first two propositions is obtained by uniting an assertion on compactness proved in Section 6 with an assertion on connectedness proved in Section 7. The third proposition is Theorem 7.6. The three propositions together assert that a continuous image in R of a closed interval is itself a closed interval. This is just another way of stating the main theorem.

We have done much more than prove our main theorem; we have proved a number of theorems of considerable generality, and we have analyzed the argument so that theorems similar to the main theorem can be obtained without further trouble. For example, the fact that a closed and bounded set is compact (Section 6), together with propositions 2 and 3, enables us to conclude:

If X is a closed, bounded, and connected set in R^n, and if $f: X \to R$ is continuous, then the image fX is a closed interval.

One of the many different kinds of closed, bounded, and connected subsets of R^n is the surface of a sphere in R^3. Hence a continuous real-valued function defined on a sphere has a maximum value, a minimum value, and takes on all values between. We can see now that the hypothesis of the main theorem, that the domain of f is a closed interval, is unnecessarily restrictive; it is enough to require that the domain of f be closed, bounded, and connected.

We are now in a position to begin to answer the question: What is topology?

DEFINITION. A property of a subset X of R^m is called a *topological* property if it is equivalent to a property whose definition uses only the notion of open set of X and the standard concepts of set theory (element, subset, complement, union, intersection, finite, infinite, etc.).

Briefly, a topological property of $X \subset R^m$ is one which is expressible as a property of the family of open sets of X.

Compactness and connectedness are topological properties. The reader is urged to review carefully the definitions of these concepts given in Section 6 and Section 7, and to note the absence of references to properties of X such as size, shape, length, area, and volume. Similarly, a closed set in X is a topological concept because a closed set is defined to be the complement of an open set in X.

Once a property or concept is known to be topological, we may use it freely in defining other topological properties and concepts; for example the notions of closed set, compactness, and connectedness may be so used.

Here are some examples of topological properties of specific sets. A line L is a connected set, and the complement in L of each point of L is not connected. Stated differently: a line is disconnected by the deletion of any one of its points. A circle does not have this property; however it is disconnected by the deletion of any pair of its points. Some of the topological properties of a plane are that it is connected, it is not compact, and it is not disconnected by the deletion of any finite set of points.

If a property of a subset X of R^m involves features of X or its subsets such as size, shape, angle, length, area or volume, then it is not likely to be a topological property. For example, the property of being bounded refers to the size of X. The property of being closed in R^m refers to the set $R^m - X$ and not just to open sets of X. On the face of it, neither property is likely to be a topological property of X. However there is danger here of jumping to a hasty conclusion. If we consider the property of X of being both bounded and closed in R^m, we might equally well presume that this is not a topological property; but we proved in Section 6 that it is equivalent to compactness which is a topological property. Clearly, we need a practical test for a property to be *not* topological. Such a test is based on the concept of the topological equivalence of two point sets:

DEFINITION. A set $X \subset R^m$ and a set $Y \subset R^n$ are called *topologically equivalent* (or *homeomorphic*) if there is a one-to-one function $f: X \to Y$ such that f is continuous and also $f^{-1}: Y \to X$ is continuous; moreover, the function f is called a *topological equivalence* (or *homeomorphism*).

Let us consider some examples. We have already observed that any two line segments are similar, and that a similarity is continuous. Since the inverse function of a similarity is also a similarity, it follows that any two line segments are topologically equivalent (see Fig. 8.1).

As shown in Fig. 8.2, we may use a radial projection with center z to define a topological equivalence between a line segment and an arc of a circle. The continuity of f is proved by forming the wedge-shaped region determined by $N(fx, \epsilon)$ and the point z, and then taking a positive δ so small that $N(x, \delta)$ lies in this region. Since f^{-1} contracts distances, it too is continuous.

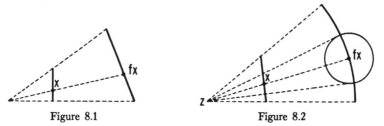

Figure 8.1   Figure 8.2

In fact a very wiggly curve can be topologically equivalent to a line segment. Fig. 8.3 shows the graph C of a continuous function f defined on an interval I. Let $g: C \to I$ be the perpendicular projection, so that $g(x, fx) = x$. Clearly g is 1–1, and $g^{-1}x = (x, fx)$ for all $x \in I$. As a projection, g contracts distances, so g is continuous. The continuity of g^{-1} is a consequence of the continuity of f. So, for any continuous f, the graph C of f is topologically equivalent to the line segment I.

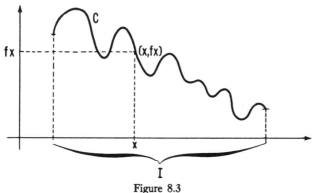

Figure 8.3

Speaking roughly, any non-selfintersecting curve described by the continuous motion of a particle moving from a point p to a point q is topologically equivalent to a line segment (see Fig. 8.4).

Figure 8.4

Another example of a homeomorphism is provided by the stereographic projection of a spherical surface S, with a pole p deleted, onto its equatorial plane P (see Fig. 8.5). The solution of Exercise 7 of Section 3 shows that this defines a topological equivalence between $S - p$ and P. If we restrict the projection to the points of a single great circle C through p, we obtain a topological equivalence of $C - p$ with a line L.

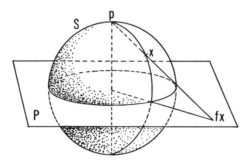

Figure 8.5

Having illustrated the concept of topological equivalence, let us return to the discussion of topological properties. The following theorem states the basic relationship involving these concepts.

THEOREM 8.1. *If a subset X of R^m and a subset Y of R^n are topologically equivalent, then each has every topological property possessed by the other.*

This is obvious because a topological equivalence $f: X \to Y$ sets up a 1-1 correspondence between the points of the two sets, and a 1-1 correspondence between their subsets ($A \subset X$ is associated with $fA \subset Y$ and $B \subset Y$ with $f^{-1}B \subset X$) in such a way as to make open sets correspond to open sets, and to preserve the relations and operations of set theory (for example, $A \subset B$ in X if and only if $fA \subset fB$ in Y). Any true statement we can make about points of X, subsets of X, open sets of X, and their set-theoretic relations, will yield a true statement if we replace all points and subsets of X by their images in Y.

Let us illustrate the argument with the property of being not connected. In terms of open sets, this is stated: X has two non-empty open sets A and B such that $A \cup B = X$ and $A \cap B = \emptyset$. If we take images under f and use the obvious relations $fX = Y$, $f\emptyset = \emptyset$, $f(A \cup B) = fA \cup fB$ and $f(A \cap B) = fA \cap fB$, then we obtain: Y has two non-empty open sets fA and fB such that $fA \cup fB = Y$ and $fA \cap fB = \emptyset$. Therefore Y is not connected.

We use Theorem 8.1 to show that certain properties of a subset X of R^m are not topological. Since a line segment is topologically equivalent to any other line segment, its length is not a topological property. Since a line segment is equivalent to an arc of a circle, its straightness is not a topological property. Since a sphere S with a point p deleted is equivalent to a plane P under stereographic projection, the boundedness of $S - p$ is not a topological property. Since P is closed in R^3 and $S - p$ is not closed in R^3, the property of P being closed in R^3 is not topological.

By this time the answer to the question "What is topology?" should be fairly obvious:

Topology is the study of the topological properties of point sets.

This is a satisfactory answer but it is not the complete answer. We must include also the topological properties of functions. If $f: X \to Y$ is a function such that $X \subset R^m$ and $Y \subset R^n$, then a property of f is called *topological* if it is equivalent to one whose definition uses only the notions: open sets of X and of Y, images and inverse images, and the standard concepts of set theory.

For example, continuity is a topological property of a function because Theorem 4.5 states that f is continuous if and only if the inverse image of each open set of Y is an open set of X. It is easy to find topological properties of functions. As another example, the property of being a constant function is a topological property. Again, the property of $f: X \to Y$ that $f^{-1}y$ is a compact subset of X for every $y \in Y$ is a topological property.

The complete answer to the original question is that *topology is the study of the topological properties of point sets and functions.*

In the light of these definitions, let us reexamine our proof of the main theorem as broken down into the three propositions given at the beginning of this section. The second proposition is pure topology; it states that a topological property of X (compactness) and a topological property of f (continuity) imply a topological property of fX (compactness). The same holds with connectedness in place of compactness. The first proposition states two topological properties of a familiar object. The third is a converse of the first: the two topological properties, compactness and connectedness, characterize closed intervals among all subsets of R. We can conclude from this that the proof of the main theorem is nearly all topological.

Topology has been called *rubber geometry*. If one attempts to picture those point sets which are topologically equivalent to a particular set X, it is a good intuitive device to regard X as made of rubber. If X can be deformed into a set Y by stretching here, contracting there, and twisting (but never tearing or gluing different parts together), then X and Y are topologically equivalent. For example, a small spherical surface (balloon) can be inflated into a big one, then it can be squeezed to form an ellipsoid, and then it can be squeezed still more to yield a surface of a dumbbell. Also, an inflated spherical surface can be allowed to contract until its surface fits tautly over the surface of a solid such as a rectangular box or a tetrahedron (Fig. 8.6).

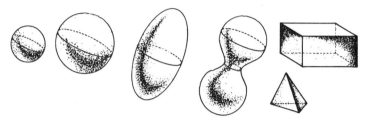

Figure 8.6

Since two topologically equivalent point sets have exactly the same topological properties, the topologist regards them as being essentially the same (topologically indistinguishable). This is analogous to the viewpoint in euclidean geometry that two congruent configurations are completely equivalent. A *topologist* has been defined to be a mathematician who can't tell the difference between a donut and a cup of coffee. Fig. 8.7 shows several intermediate stages of the deformation of a solid donut into a cup.

Figure 8.7

In each of the classical geometries there is a concept of equivalent configurations. As already noted, in euclidean geometry two configurations are equivalent if they are congruent, in particular, if there is a rigid motion carrying one onto the other. In projective geometry, figures are equivalent if there is a *projectivity* carrying one onto the other. The projectivities include congruences and similarities, and enough additional transformations so that any two triangles are equivalent, and any circle is equivalent to any ellipse (see Fig. 8.8).

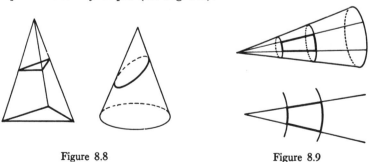

Figure 8.8   Figure 8.9

In differential geometry, the equivalences are called isometries (equal metrics). Here, two configurations are equivalent if there is a 1–1 correspondence between their points such that the length of any curve in the domain equals the length of the corresponding curve in the range. For example, a portion of a cylindrical surface can be rolled out upon a portion of a plane. Likewise a portion of a cone can be rolled out onto a sector of an annular ring (see Fig. 8.9). Therefore these surfaces are isometric.

Congruences, projectivities, and isometries are all topological equivalences; for, in each case, the correspondence between two equivalent figures is 1–1 and continuous each way. It follows that every topological property of one such configuration is also a property of the other, and therefore a topological property is also a property in the sense of euclidean geometry, projective geometry, and differential geometry. As a consequence, a theorem of topology is automatically a theorem of each of these geometries. In view of this, it can be said with considerable justification that topology is the fundamental geometry.

Exercises

1. Find a homeomorphism between X and Y if:

   (a) X is an open interval and Y is a line.

   (b) X is a half-open interval and Y is a ray.

   (c) X is the interior of a circle and Y is the plane.

2. Show that a circle with one point deleted is topologically equivalent to an open interval.

3. A set is totally disconnected if the only connected subsets are single points and the empty set. Give two examples of subsets of R that are totally disconnected and contain infinitely many points.

4. A set X is locally connected if, for each point $x \in X$ and each neighborhood $N(x, r)$, there is a connected open set U of X such that $x \in U \subset N(x, r)$. Determine whether each of the following sets is locally connected:

   (a) the set of all integers,   (b) the set $\{0, \frac{1}{1}, \frac{1}{2}, \frac{1}{3}, \cdots\}$,

   (c) the set $[a, b]$,   (d) the set of rational numbers.

5. A set X is locally compact if each point of X has a neighborhood in X contained in a compact subset of X.

   (a) Give an example of a set that is not compact but is locally compact.

(b) Show that each closed set in R^m is locally compact.

6. Determine which of the following properties are topological and which are not. For each property given that is not topological, find two topologically equivalent sets, one with the property and one without.

    (a) X is unbounded.
    (b) X is a finite set.
    (c) X is a curve of length 2.
    (d) X is locally compact.
    (e) X is a convex polygon.
    (f) X is locally connected.
    (g) X is totally disconnected.

7. Illustrate the argument of Theorem 8.1 to show that, if a compact subset X of R^m and a subset Y of R^n are topologically equivalent, then Y is compact.

8. Which of the following properties of a function $f: X \to Y$ are topological?

    (a) The image of each open set of X is an open set of Y.
    (b) f is a similarity.
    (c) f is a translation.
    (d) The inverse image of each point is a finite set.
    (e) The inverse image of each point is a compact set.
    (f) The inverse image of Y is bounded.
    (g) The inverse image of each point is a connected set.

9. A fixed point theorem

If a set is mapped into itself by a function f, it may happen that some point is carried into itself. A point x with the property that $fx = x$ is called a *fixed point of the mapping*. If a circular disk is rotated on itself through a right angle, the center of the disk is the sole fixed point. The same mapping restricted to the circle which is the periphery of the disk has no fixed point. Each constant map of a space into itself has one fixed point. Thus a mapping of a set into itself may or may not have a fixed point depending on the set and the mapping. However, in the case of a line segment (closed interval), we have the following remarkable result.

THEOREM 9.1. *Every mapping of a line segment into itself has at least one fixed point.*

Let coordinates be introduced on the line so that the segment becomes an interval $[a, b]$. Then a mapping of the segment into itself is just a continuous function $f: [a, b] \to [a, b]$. Define a new function $g: [a, b] \to R$ by $gx = fx - x$ for each $x \in [a, b]$. Thus g measures the directed distance between x and its image fx. It is positive when fx is to the right of x, i.e. $fx > x$, and negative when fx is to the left of x. We seek a fixed point of f, that is, a point at which g is zero. If either endpoint is fixed, we have nothing to prove. Suppose neither is fixed. Since fa and fb are in $[a, b]$, $a < fa$ and $fb < b$; hence $ga > 0$ and $gb < 0$. Since g is continuous (it is the difference of two continuous functions, see Exercise 8 of Section 3), the main theorem asserts that g takes on all values between ga and gb. So $gx = 0$ for some $x \in [a, b]$, and this x is the required fixed point of f.

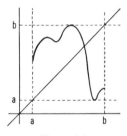

Figure 9.1

Theorem 9.1 may be examined from the point of view of the graph of f illustrated in Fig. 9.1. A fixed point of f is one whose corresponding point on the graph lies on the diagonal line (i.e. if $fx = x$, then $(x, fx) = (x, x)$ is on the diagonal). Since $a < fa$, the point (a, fa) lies above the diagonal, and similarly (b, fb) lies below it. Since the diagonal line disconnects the plane into the points above and those below, and the graph is a connected set, the graph must intersect the diagonal. The function g measures the vertical distance between the graph and the diagonal.

Exercises

1. Find the fixed point of the mapping of the interval $[0, 1]$ onto itself defined by $fx = (1 - x^2)^{1/2}$.

2. Let the mapping of the interval $[0, 1]$ onto itself be defined by $fx = 4x - 4x^2$.

    (a) Sketch the graph of the function and the diagonal line $y = x$.

    (b) Is the mapping $1 - 1$ in the interval?

    (c) Find the fixed points of the mapping.

3. Let the mapping of the interval $[0, 1]$ into itself be defined by $fx = x^2 - x + 1$.

   (a) Sketch the graph of the function and the line $y = x$.

   (b) Is this mapping $1-1$ in the interval?

   (c) Find the fixed points of the mapping.

4. Show that the following property of a set X in R^n is a topological property: every mapping of X into itself has a fixed point.

5. Give an example of a mapping of the interval $[0, 1]$ into itself having precisely two fixed points, namely 0 and 1.

6. Give an example of a mapping of the open interval $(0, 1)$ *onto* itself having no fixed points.

7 Show that each mapping of a half-open interval *onto* itself has at least one fixed point.

10. Mappings of a circle into a line

A circle has the following striking property:

THEOREM 10.1. *Every mapping of a circle into a line carries some pair of diametrical points into the same image point.*

PROOF. Let $f\colon C \to L$ be a mapping of a circle C into a line L. By introducing coordinates on L, we may consider f as having the real numbers R as its range. Consider a pair of diametrical points p and p' on C (Fig. 10.1); let their image points on L have coordinates $fp = a$ and $fp' = b$, and examine the function g defined by

$$gp = fp - fp' = a - b.$$

This is a continuous function of p because f is continuous. Moreover,

$$gp' = fp' - fp = b - a = -(a - b),$$

so the function g is either zero at p and at p' (in which case p and p' have the same image under f), or it has opposite signs at p and at p'. In the second case, we apply the main theorem to one of the semicircles from p to p' to obtain a point q such that $gq = 0 = fq - fq'$. It follows that $fq = fq'$, that is, the diametrical points q and q' have the same image point.

The analog of diametrical points on a circle is *antipodal points* on an ellipse; these are points located symmetrically with respect to the center of an ellipse. Since a circle is a special case of an ellipse, diametrical is a special case of antipodal. It is therefore appropriate to ask whether a similar theorem holds for antipodal points.

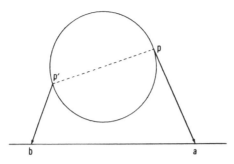

Figure 10.1

If a circle X and an ellipse Y have the same center z and lie in the same plane, a homeomorphism is most easily constructed by pairing off two points if they lie on the same ray from z. This is essentially the radial projection onto X mentioned in Section 3 and has been proved continuous in that section. If X and Y are not concentric, we arrive at a homeomorphism by first projecting Y radially onto a circle X' that has the same center as Y (see Fig. 10.2). Since X and X' are similar, X is topologically equivalent to X' which is in turn topologically equivalent to Y. Furthermore, the homeomorphism $Y \to X$, composed of the radial projection and the similarity, preserves antipodes; that is, if the image of q is p, and if q' is the antipode of q, then the image of q' is the antipode p' of p. So the theorem on diametrical points holds for the ellipse with the understanding that antipodal points play the same role as diametrical points.

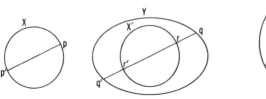 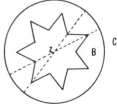

Figure 10.2 Figure 10.3

A similar argument shows that the result holds also for any star-shaped closed curve such as the polygon B in Fig. 10.3. By projecting B radially from the center point z onto the circle C, we obtain a homeomorphism which transforms each pair of points of B on the same line through z into a diametrical pair on C.

Exercises

1. In Fig. 10.4, let $f: C \to L$ be the projection of the circle into the line from a point p outside C as the center of the projection.

   (a) Describe the inverse images of various points of L.

   (b) Which pair of diametrical points on C have the same image in L?

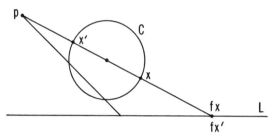

Figure 10.4

2. Let L be tangent to a circle C at p. From the point p' of C diametrical to p, project C onto L. Describe the inverse images of points of L. Why doesn't Theorem 10.1 apply to this mapping?

3. If a circle C is divided by diametrical points b and b' into two semicircles D and D', show that, in any mapping $f: D \to D'$, some point is reflected about the diameter bb'.

4. If a circle C is divided by diametrical points b and b' into two semicircles D and D', show that any mapping $f: D \to D'$, carries some point into its antipode.

5. Give an example of a non-constant map of the circle into the line such that each two diametrical points have one image point.

11. The pancake problems

The first pancake problem may be stated roughly as follows: Suppose two irregularly shaped pancakes lie on the same platter; show that it is possible to cut both exactly in half with one stroke of the knife. If, for example, each pancake happens to form a perfect circle, then the line through their centers would provide the desired cut. The problem becomes more difficult, however, if the shapes of the pancakes are not restricted. The precise mathematical theorem is as follows.

PANCAKE PROBLEMS

THEOREM 11.1. *If A and B are two bounded regions in the same plane, then there is a line in the plane which divides each region in half by area.*

By a region in the plane is meant a connected open subset of the plane. The theorem applies even when the two pancakes are stacked one onto the other, that is, the two regions may overlap.

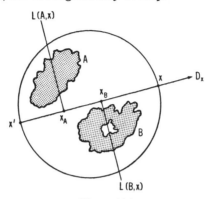

Figure 11.1

Since the proof is somewhat long, we present first its major steps, deferring the proofs of two minor propositions. Since A and B are bounded, we may choose a circle C including $A \cup B$ in its interior (see Fig. 11.1). Let z denote the center of C and r its radius. For any $x \in C$, let x' denote its diametrical point, and D_x the diameter x' to x. Our first proposition to be proved later is

(1) For any $x \in C$, the family of all lines perpendicular to D_x contains one and only one line $L(A, x)$ which divides A in half by area, and one and only one line $L(B, x)$ which divides B in half by area.

Denote by x_A and x_B the points where D_x meets $L(A, x)$ and $L(B, x)$, respectively. On D_x we have a natural scale (or coordinate system) with z at the origin: the coordinate of a point is its directed distance from z, positive when the point is on the same side as x, negative otherwise. Let $g_A x$ and $g_B x$ denote the coordinates of x_A and x_B, respectively. Now, for each $x \in C$, set

$$hx = g_A x - g_B x.$$

Our second proposition to be proved later is

(2) The function $h: C \to R$ is continuous.

A crucial property of h is that its values at any two diametrical points of C are equal in absolute value but opposite in sign:

$$hx' = -hx \qquad \text{for any } x \in C.$$

This is proved by noting that $D_{x'} = D_x$, so $L(A,x) = L(A,x')$ and $L(B,x) = L(B,x')$; hence $x'_A = x_A$ and $x'_B = x_B$. However, the positive direction for coordinates on $D_{x'}$ is opposite that of D_x; hence $g_A x' = -g_A x$ and $g_B x' = -g_B x$, and therefore

$$hx' = g_A x' - g_B x' = -g_A x + g_B x = -hx.$$

Now, by Theorem 10.1, there is a point x of C such that $hx' = hx$. For this x we have both $hx' = hx$ and $hx' = -hx$; hence $hx = 0$, and this implies $x_A = x_B$, so that $L(A,x) = L(B,x)$ divides both A and B in half by area.

PROOF OF (1): Corresponding to a number y, let L_y denote the line perpendicular to D_x through the point of D_x with coordinate y, and let fy denote the area of the part of A on the positive side of L_y (the side in the direction of increasing y-values). Then f is a real-valued function of a real variable, $f: R \to R$. As y varies from $-r$ to r, the line L_y sweeps once over the interior of C. Picture L_y as a steel needle mounted on a rod D_x at right angles. As the mounting traverses the rod from x' to x, the needle sweeps across the interior of C. When $y = -r$, the mounting is at x', all of A is on the positive side, so $f(-r)$ is the area of A. When $y = r$, the mounting is at x, all of A is on the negative side, so $fr = 0$.

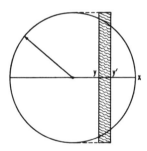

Figure 11.2

To show that f is continuous, let $y, y' \in R$ with $y < y'$. Then $fy - fy'$ is the area of the part of A between the lines L_y and $L_{y'}$. Since this is contained in the rectangular region shown as shaded in Fig. 11.2, it follows that $|fy - fy'| < 2r|y - y'|$. Corresponding to an $\epsilon > 0$, we take $\delta = \epsilon/2r$. Then, when y' is in the δ-neighborhood of y, it follows that fy' is in the ϵ-neighborhood of fy. Therefore f is continuous at y. Since this is true for each y, f is continuous.

By the main theorem, as y varies from $-r$ to $+r$, fy sweeps over all values starting from the area of A down to 0. Hence there is at least one y-value where fy is exactly half the area of A so that $L_y = L(A,x)$ cuts A in half. We need to know that there is only one such cut. Suppose to the contrary that both L_y and $L_{y'}$ divide A in half (i.e. $fy = fy'$) and that $y \neq y'$, say $y < y'$. The strip Q between L_y and $L_{y'}$ is an open set, and its complement separates into two parts, one containing the positive side of $L_{y'}$ and the other the negative side of L_y. Since A is connected and contains points in each of the two parts, A must contain a point of Q, say p. Since A and Q are open, $A \cap Q$ is open, so it contains a neighborhood of p. Therefore $A \cap Q$ has positive area, hence $fy > fy'$. Since this contradicts $fy = fy'$, we have proved the uniqueness. The existence and uniqueness of $L(B, x)$ is proved similarly. This completes the proof of (1).

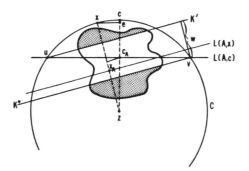

Figure 11.3

PROOF OF (2): Since h is the difference $g_A - g_B$, it suffices to prove that g_A and g_B are continuous (see Exercise 8 of Section 3). Let c be a point of C where we wish to show that g_A is continuous, and, in accord with the notation above, let c_A be the point on the diameter D_c where the perpendicular $L(A, c)$ cuts A in half (see Fig. 11.3). Let x be a point of C near c. Through the points u and v where $L(A, c)$ meets C, draw lines K and K' perpendicular to D_x. The line $L(A, c)$ divides the interior of C into two parts, U and V. The strip between K and K' separates its complement in the interior of C into two parts, U' and V', such that $U' \subset U$ and $V' \subset V$. Therefore U' and V' each contain at most half the area of A. It follows that the line $L(A, x)$, perpendicular to D_x and dividing A in half, lies in the strip, and so does the point x_A where $L(A, x)$ meets D_x. Since the circle through c_A with center z meets D_x inside the strip, it follows that

$$|g_A x - g_A c| < w,$$

where w is the width of the strip.

To obtain an estimate of the size of w, notice that the similarity of two triangles gives

$$\frac{w}{d(u, v)} = \frac{d(e, x)}{d(z, x)},$$

where e is the foot of the perpendicular from x to D_c. Since $r = d(z, x)$, this gives

$$w = \frac{d(u, v)}{r} d(e, x).$$

Since $d(u, v) \leq 2r$, and $d(e, x) \leq d(c, x)$, we obtain

$$w \leq 2d(c, x),$$

and therefore

$$|g_A x - g_A c| \leq 2d(c, x).$$

So if $\epsilon > 0$, and $x \in N(c, \epsilon/2)$, it follows that

$$|g_A x - g_A c| < \epsilon.$$

This shows that g_A is continuous. Similarly g_B is continuous. This completes the proof of (2) and Theorem 11.1.

For our second pancake problem we are asked to cut one pancake into four equal parts with two perpendicular cuts.

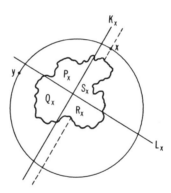

Figure 11.4

THEOREM 11.2. *If A is a bounded region in the plane, then there are two perpendicular lines which divide A into four parts having equal areas.*

As before we enclose A within a circle C. For each $x \in C$, let L_x be the line perpendicular to D_x which divides A in half, and let K_x be the line parallel to D_x which divides A in half. The two lines divide A into four parts which, counting in a counter-clockwise direction (see Fig. 11.4), have areas denoted by P_x, Q_x, R_x, S_x. Since L_x and K_x divide A in half, we have

$$P_x + Q_x = R_x + S_x \quad \text{and} \quad Q_x + R_x = S_x + P_x.$$

§11] PANCAKE PROBLEMS 69

Subtracting these equations, we obtain $P_x = R_x$ and $Q_x = S_x$. If by good fortune we also have $P_x = Q_x$, then the lines L_x and K_x, solve our problem. Since this is generally not the case, we set the difference $P_x - Q_x = fx$ and ask how this function varies as x moves around the circle. If $y \in C$ is such that D_y is perpendicular to D_x, it is clear that $L_y = K_x$ and $K_y = L_x$. It follows that $P_y = Q_x$ and $Q_y = R_x$. Since $P_x = R_x$, we obtain

$$fy = P_y - Q_y = Q_x - P_x = -(P_x - Q_x) = -fx.$$

Therefore the function f reverses sign as x moves through an arc of 90°. Once f has been shown to be continuous, it will follow from the main theorem that $fx = 0$ somewhere on each arc of 90°. Such a point provides the required dissection.

We shall only sketch the proof of continuity. Since f is the difference of two functions, it will again suffice to show that P_x is continuous (a similar proof shows that Q_x is continuous). Let $c \in C$ be a point where the continuity of P_x is to be proved, and let x be near to c. The passage from the perpendicular pair L_c, K_c to the similar pair L_x, K_x can be done in two steps. First rotate L_c, K_c about their point p of intersection into a pair of perpendiculars L'_c, K'_c which are parallel respectively to L_x, K_x. The angle α of rotation is the small one determined by the arc c to x. The second step translates L'_c, K'_c into L_x, K_x by a parallel displacement. The change from P_c to P'_c is seen to be no more than the area of the off-center sector of C with vertex p and angle α, and this is at most $2r\,d(c,x)$ where r is the radius of C. The area U between L'_c and L_x and inside C is at most $2ru$ where u is the distance between L'_c and L_x. Similarly the area V between K'_c and K_x is at most $2rv$. The change from P'_c to P_x is clearly less than

$$U + V \leq 2r(u + v).$$

The point q in which L_c and L_x intersect can be seen to lie inside C because L_c and L_x divide A in half and A is connected. This means that $d(p, q) < 2r$, and then, by similar triangles, $u < 2d(c, x)$. In the same way, $v < 2d(c, x)$. Putting these estimates together gives

$$|P_x - P_c| < 10r\,d(c, x).$$

So, if a number $\epsilon > 0$ is given, we take $\delta = \epsilon/10r$, and then we have $|P_x - P_c| < \epsilon$ for every $x \in N(c, \delta)$. This concludes the proof.

Exercises

1. Two pancakes, one in the shape of a perfect square and the other in the shape of a perfect circle lie on the same platter. Describe the cut dividing each exactly in half with one stroke of the knife.

2. Would the "line of centers" method work for any two pancakes in the shape of regular polygons?

3. In how many ways can cuts be made to divide a square pancake into four equal parts with a set of two perpendicular lines?

4. In dividing any pancake into four equal parts with a pair of perpendicular lines, the function $P_x - Q_x$ is zero within each quarter-turn. Explain why this does not necessarily imply at least four such divisions as x traverses the entire circle.

5. If one pancake is circular and the other is irregular, give a direct argument (different from that in the text) to show that some single line cuts both in half.

6. In Theorem 11.1, replace the knife which makes straight cuts by a blade in the shape of a semicircle having a radius equal to the diameter of the circle enclosing the two regions, and, in the analog of proposition (1), consider those cuts for which the center of the semicircle is on the ray from z through x. Where does the argument fail for this type of cut? For what type of curved blade would the argument hold?

12. Zeros of polynomials

Our next theorem is an application of the main theorem to algebra.

THEOREM 12.1. *A polynomial of odd degree with real coefficients has at least one real zero.*

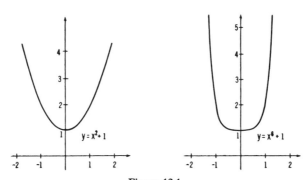

Figure 12.1

To appreciate the implications of this theorem, let us examine some specific examples of polynomials of even degree and polynomials of odd degree. First, if the polynomial has degree 1, $fx = ax + b$, $a \neq 0$, then the graph of $y = ax + b$ is a line that crosses the x-axis at $x = -b/a$, so the polynomial has a zero for this value of x. Next, consider the parabola $y = x^2 + 1$ as an example of a polynomial of degree 2 (see Fig. 12.1). The curve lies entirely in the upper half of the coordinate plane. The minimum value of $x^2 + 1$ is 1 because for any real number x, $x^2 \geq 0$; hence the polynomial has no real zero. Similarly, $x^6 + 1$ has no real zero; neither has $x^4 - 2x^2 + 5$ since

$$x^4 - 2x^2 + 5 = (x^2 - 1)^2 + 4$$

never has a value less than 4 (see Fig. 12.2). On the other hand the polynomial $x^2 - 4x + 3$ of even degree has the zeros $x = 1$ and $x = 3$ (see Fig. 12.3).

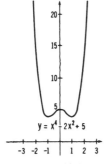

Figure 12.2

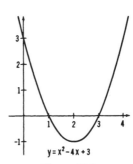

Figure 12.3

The graph of $y = x^3 - x + 5$ is the curve shown in Fig. 12.4; it crosses the x-axis somewhere between -2 and -1. The polynomial $x^5 - 2x^3 + x + 4$ has degree 5; its graph, sketched in Fig. 12.5, crosses the x-axis somewhere between -1.7 and -1.6.

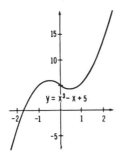

Figure 12.4

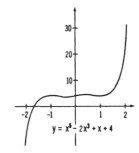

Figure 12.5

In each of our examples, the graph of an odd degree polynomial rises from $-\infty$, crosses the x-axis, and eventually goes on to $+\infty$. Even degree polynomials have graphs that come down from $+\infty$ and go back to $+\infty$ with a few possible wiggles between, and our examples show that some of these never cross the x-axis. The gist of Theorem 12.1 is that this is not the case with polynomials of odd degree; every odd degree polynomial with real coefficients has at least one real zero.

To prove Theorem 12.1, it suffices to consider polynomials of the form

$$f(x) = x^n + a_1 x^{n-1} + \cdots + a_{n-1} x + a_n,$$

for, if the coefficient of the term of highest degree is not 1, we can multiply the polynomial by the reciprocal of this coefficient without changing its zeros. For $x \neq 0$, we may write $f(x)$ in the form

$$x^n \left(1 + \frac{a_1}{x} + \cdots + \frac{a_{n-1}}{x^{n-1}} + \frac{a_n}{x^n} \right),$$

or $f(x) = x^n q(x)$, where

$$q(x) = 1 + \frac{a_1}{x} + \cdots + \frac{a_{n-1}}{x^{n-1}} + \frac{a_n}{x^n}.$$

Our method of proof will consist in showing that the polynomial f, of odd degree, is negative for some x, positive for some other x, and continuous. The main theorem will then yield the desired result.

Now if x is a number such that the absolute value of each of the terms

$$\frac{a_1}{x}, \frac{a_2}{x^2}, \cdots, \frac{a_n}{x^n}$$

is less than $1/n$, then the sum

$$h(x) = \frac{a_1}{x} + \frac{a_2}{x^2} + \cdots + \frac{a_n}{x^n}$$

of these n terms is less in absolute value than $n/n = 1$; this means that $h(x)$ is between -1 and $+1$, and since $q(x) = 1 + h(x)$, $q(x)$ is positive. To find an x for which this holds, examine each of the numbers

$$n|a_1|, \ (n|a_2|)^{1/2}, \ \cdots, \ (n|a_n|)^{1/n},$$

and choose a number b greater than all of them. To see that $q(x)$ is positive for an x such that $|x| \geq b$, we observe that the inequalities,

$$|x| > n|a_1|, \ |x| > (n|a_2|)^{1/2}, \ \cdots, \ |x| > (n|a_n|)^{1/n}$$

imply

$$\left|\frac{a_1}{x}\right| < \frac{1}{n}, \ \left|\frac{a_2}{x^2}\right| < \frac{1}{n}, \ \cdots, \ \left|\frac{a_n}{x^n}\right| < \frac{1}{n}.$$

For values of x such that $|x| \geq b$ the sign of the polynomial is the sign of x^n because $f(x) = x^n q(x)$ and $q(x)$ is positive. Since n is odd, x^n has the sign of x. Thus the polynomial is positive for $x = b$ and negative for $x = -b$.

To apply the main theorem and infer the existence of a zero between $-b$ and b, it is necessary to show that a polynomial is a continuous function. In Section 3 we showed that any constant function (e.g., a polynomial of degree 0) and any identity function (e.g., the polynomial x of degree 1) are continuous. In Exercise 2 below, it is required to prove that a product of continuous functions is continuous. From this it follows that $x^2 = x \cdot x$ is continuous, $x^3 = x^2 \cdot x$ is continuous and, by an induction, x^k is continuous for every k. Since x^k and a constant a are continuous, the same result tells us that any monomial ax^k is continuous. Now every polynomial is the sum of its monomial terms, and any sum of continuous functions is continuous (see Exercise 8, Section 3 and answer); therefore every polynomial is continuous.

Exercises

1. Show that the polynomial x^2 is a continuous function.

2. Show that the product of two continuous functions f and $g: [a, b] \to R$ is a continuous function. Hint:
$$| (fx)(gx) - (fx')(gx') | = | (fx)(gx - gx') + (fx - fx')(gx') |$$
$$\leq |fx| \, |gx - gx'| + |fx - fx'| \, |gx'|.$$

3. For a polynomial of degree n, what is the determining factor as to whether the polynomial is positive or negative when x is zero?

4. Use the criterion $|x| > (n|a_k|)^{1/k}$ to find a number b such that the polynomial $f(x) = x^3 - 2x^2 - 3x$ is positive for $x > b$ and negative for $x < -b$. Factor the polynomial into linear factors to find the smallest number a such that $f(x) > 0$ for $x > a$, and the largest number c such that $f(x) < 0$ for $x < c$.

5. Use the criterion $|x| > (n|a_k|)^{1/k}$ to find a number b such that the polynomial $x^5 - 3x^4 + 125x^3 + 200x^2 - x + 2$ is positive for $x > b$ and negative for $x < -b$. (Notice that the cubic in Exercise 4 was easy to factor into linear factors, but this is not so for the quintic of this problem.)

PART II

Existence Theorems in Dimension 2

13. Mappings of a plane into itself

As stated in the introduction, our purpose in Part II is to prove an existence theorem for solutions of pairs of simultaneous equations. This theorem is stated in Section 18, and its proof is completed in Section 26. Sections 27 through 36 apply the theorem to questions about fixed points of mappings, singularities of vector fields, and zeros of polynomials. To formulate the main theorem, we must develop two-dimensional analogs of the one-dimensional concepts of Part I. The crucial concept needed is that of the winding number of a closed curve in a plane about some point of the plane not on the curve. We shall give first an intuitive definition of this notion together with an intuitive proof of the main theorem (Sections 17, 18). In Sections 19–26, the definition is made precise and the proof rigorous.

Recall that the main theorem of Part I deals with a mapping $f: [a, b] \to R$ of a segment into a line, and gives conditions on a point $y \in R$ under which it could be asserted that y is in the image $f[a, b]$ (e.g. $fa \leq y \leq fb$). The main theorem of Part II will deal with a mapping $f: D \to P$ of a portion D of a plane $P (= R^2)$ into P, and will give conditions on a point $y \in P$ under which it can be asserted that y is in the image fD.†

† In the introduction this theorem was described as dealing with a pair of simultaneous equations $f(x, y) = a$ and $g(x, y) = b$; this form is converted to the present form by making the notational substitutions (x_1, x_2) for (x, y), (y_1, y_2) for (a, b), f for (f, g), and by interpreting the pairs of numbers (x_1, x_2), (y_1, y_2) as coordinates of points x, y in P.

Recall also that we found the concept of the graph of $f: [a, b] \to R$ very useful in explaining the meaning of the main theorem and in making its truth geometrically evident. In the two-dimensional case, we may also speak of the graph of a mapping $f: D \to P$. To see what it involves, note that a point of the plane $P = R^2$ is represented by two real numbers (x_1, x_2). Its image under f requires two more, (y_1, y_2). Then the pair consisting of the point and its image is represented by *four* numbers, and a point of the graph is a point of four-dimensional space. Thus the graph of f is a curved surface in R^4.

Here then is our first difficulty. To explain our theorems by graphs would require the ability (which none of us has) to visualize a surface in four dimensions. We must therefore adopt a different method of visualizing mappings: the method of picturing images and inverse images as described briefly in Section 2 of Part I. In the remainder of this section, we shall discuss more complicated mappings by the same method. Our purpose is to sharpen the geometric intuition, and to indicate the degree of generality of subsequent theorems.

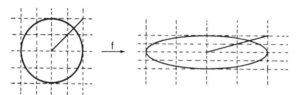

Figure 13.1

In Part I, Section 2, we discussed translations, rotations, reflections and similarities as mappings of the plane into itself. A more complicated mapping is one which expands lengths in one direction and contracts them in another. Fig. 13.1 illustrates a mapping f which doubles lengths in the horizontal direction and halves lengths in the vertical direction. Clearly it alters angles and shapes. It maps a circle into an ellipse. Surprisingly it maps any straight line into a straight line.

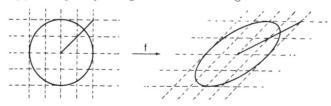

Figure 13.2

Fig. 13.2 illustrates a shearing transformation $P \to P$. Picture a trellis with many horizontal and vertical slats with a nail inserted at each junction of a horizontal and a vertical slat. Such a structure is not rigid and can slew over, exercising a scissoring action on unwary fingers. A shear also maps circles into ellipses and straight lines into straight lines.

§13] MAPPINGS OF PLANE INTO ITSELF 77

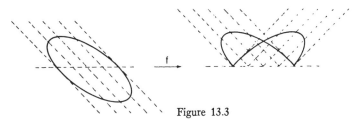

Figure 13.3

The mappings considered so far are 1 to 1 mappings. We want to consider also mappings which are not 1 to 1. Fig. 13.3 illustrates a simple fold of P about a line. This mapping is 1 to 1 along the line of the fold, but every point above the line of the fold is the image of two distinct points of the plane.

Figure 13.4

Fig. 13.4 illustrates a doubling of a plane on itself. A center point z is mapped into itself. Each point of a ray L is also fixed. Each ray issuing from z is mapped rigidly onto a ray issuing from z but forming an angle with L which is twice the original angle. Think of a ray rotating about z at a constant angular velocity; its image is a ray rotating about z at twice the velocity. As the first ray completes a half-rotation, its image completes a full rotation. This mapping is 2 to 1 except at z. Each circle with center z is wrapped twice around itself. A similar mapping is obtained for each integer n by multiplying the angular velocity by n. It is n to 1 except at z.

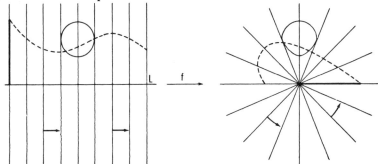

Figure 13.5

An even more complicated mapping is one which winds P over P an infinity of times. This is illustrated in Fig. 13.5. The horizontal line

L is mapped into the single point z, and each vertical line is mapped rigidly on a line through z. As the vertical line moves horizontally at a constant velocity its image spins about z at a constant angular velocity. The figure shows only a half of one rotation. This mapping is ∞ to 1. The inverse image of z is an entire ray. The inverse image of any other point consists of two rows of isolated points, one above and one below L, successive points of a row being at a fixed distance apart.

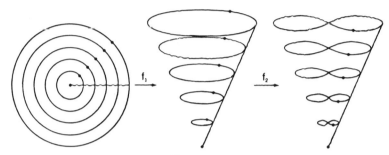

Figure 13.6

Mappings that can be described precisely and quickly are usually too simple to illustrate the complexities to be found in general. Fig. 13.6 illustrates a more complicated mapping which we shall not describe in detail. It carries a family of concentric circles into a family of figure eights. In this case the image of P is just a part of P bounded by two rays. Intuitively, this may be thought of as a stringing out of the concentric circles along a line with each circle given a half-twist.

Exercises

1. Construct a mapping $f: P \to P$ by, first, rolling up the plane P onto a cylinder Q so that the lines parallel to the x-axis are parallel to the axis of Q, and then composing this mapping with a perpendicular projection of Q onto P, assuming that the axis of Q is parallel to the x-axis. Describe the image under f of

   (a) the plane P, (b) a horizontal line $y = $ constant,

   (c) a vertical line $x = $ constant, (d) a sloping line.

   (e) Describe the inverse image of a point.

2. Let P be a plane through the center of a sphere S. Construct $f: P \to P$ as the composition of, first, the stereographic projection $P \to S$ from the pole p, and then the perpendicular projection of S back into P. Describe (a) the image of P, (b) the image of a line L in P, (c) the inverse image of a point of fP.

3. If f is the mapping of Fig. 13.5 above, describe the image of
   (a) a vertical line,
   (b) a horizontal line at a distance r from the line L,
   (c) a sloping line.
   (d) Sketch the inverse image of the configuration of Fig. 13.7, showing the first one and one-half revolutions.

Figure 13.7

14. The disk

Intervals played an important role in both the statements and proofs of our one-dimensional theorems. The analogous role for the two-dimensional theorems will be played by circular disks. A *disk* D in the plane P consists of a circle C and all points of P inside C. The circle C is called the *boundary* of D. A disk D is specified by its center point z and its radius r. A point is in the disk if its distance from z is less than or equal to the radius; that is, $x \in D$ means that $d(x, z) \leq r$.

We have seen that any two intervals are similar and therefore topologically equivalent; the same is true of any two disks D and D'. If D and D' do not have the same center we may translate D' to a disk D'' having the same center z as D. Then a suitable expansion or contraction about z will map D'' onto D.

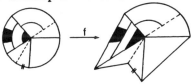

Figure 14.1

In one dimension, any subset of a line topologically equivalent to an interval is itself an interval because it must be compact and connected. In the plane, however, there are many quite dissimilar subsets which are topologically equivalent to a disk. For example, under a shearing transformation, a disk can be mapped into an ellipse and its interior. Fig. 14.1 illustrates a topological mapping of a disk into a simple closed figure and its interior. The center z is mapped into z' and each radial segment zy of D is mapped onto the parallel segment $z'y'$ by a similarity (an expansion or a contraction). The same device works also for any convex polygon, for example, a triangle or a rectangle.

In the theorems of Part II, the word disk can be interpreted as meaning any one of these configurations topologically equivalent to a disk. The disk is preferred to the others because of its symmetry and ease of description.

A disk D is certainly a bounded set since it lies inside any circle about its center having a larger radius. It is also a closed set, since each point of its complement has a neighborhood not meeting D; for, if y is not in D, then $d(y, z)$ exceeds the radius r of D, and then the circular neighborhood of y of radius $r' = d(y, z) - r$ contains no point of D. Since D is closed and bounded, it follows that D is a compact set. Hence, under any mapping $f: D \to P$, the image fD is compact, and therefore closed and bounded.

For any two points of D, the line segment joining them also lies in D. Therefore D is a connected set. It follows that fD is connected for any mapping $f: D \to P$.

Exercises

1. If a disk is cut in half by a diameter, show that half of the disk, including the diameter, is homeomorphic to a disk. Conclude that any space homeomorphic to a disk is homeomorphic to this half-disk.

2. If D is a disk and C is its bounding circle, show that any homeomorphism $g: C \to C$ can be extended to a homeomorphism $f: D \to D$.

3. If A and B are two subsets of P, both homeomorphic to a disk, and if $A \cap B$ is an arc of the boundary curve of each, show that $A \cup B$ is homeomorphic to a disk.

4. Which of the configurations in Fig. 14.2 are homeomorphic to a disk?

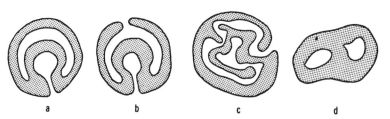

Figure 14.2

5. (a) If the last configuration of Fig. 14.2 is cut by removing a thin strip at A, as shown in Fig. 14.3, is the resulting set homeomorphic to a disk? (b) Which combinations of cuts at A, B and C will produce a homeomorph of a disk? (c) If a configuration has three holes, how many cuts are required to produce a homeomorph of a disk?

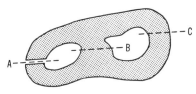

Figure 14.3

15. Initial attempts to formulate the main theorem

Our main existence theorem in two dimensions is analogous to the main theorem in one dimension. It states that, if $f\colon D \to P$ is a mapping of a disk into the plane, then an equation $fx = y$ has a solution $x \in D$ for each point y of P which satisfies a certain condition. The formulation of this condition will be a bit complicated. We shall approach it in stages by showing that several simple but plausible conditions are not adequate.

In the one-dimensional theorem, where D is a closed interval $[a, b]$, the condition on y is that it lie between fa and fb. Now a and b are the extremes of the interval, and separate it from the rest of the line. In the case of a disk, the extreme points of D are the points of the bounding circle C, and C separates D from the rest of the plane. Thus the condition to be formulated might state that y is related in some way to fC. Clearly to say "y is between fC" is nonsense. If we restate the one-dimensional condition by requiring that y be *enclosed* by fa and fb, it conveys the same idea, and the two-dimensional analogy "y is enclosed by fC" has intuitive meaning. Let us try to formulate this expression precisely. As a first attempt, consider "y is a point of the disk whose boundary is fC". This is not adequate since, for many mappings f, fC is not a circle. It could easily be an ellipse or a rectangle. As a second attempt, consider "y is a point of the region whose boundary is fC". This is better but it does not allow for an fC which is a figure 8. As a third attempt, we try "y is a point of some bounded region whose boundary is contained in fC". This seems to be what is wanted until we examine the example of an $f\colon D \to P$ in Fig. 15.1. This mapping is best described in stages pictured from left to right. First, stretch D into a long thin strip E. Next, bend E around into a curved shape F which resembles a thickened three quarters of a circle. Continue this bending until the two tips are made to overlap in the final configuration fD. The point labeled y is not in fD, yet it belongs to a bounded region whose boundary lies in fC.

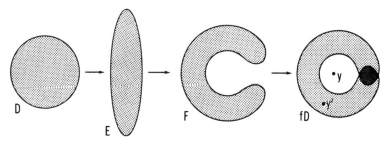

Figure 15.1

This last example exhibits clearly the difficulty we must surmount. What relationship does the point labeled y' of fD bear to fC that the point y does not bear to fC? The answer will be given in terms of a new concept we shall develop: *the winding number of a curve about a point*. We shall see that the winding number of $f \mid C$ about y' is not zero, and its winding number about y is zero. This is why $fx = y'$ has a solution $x \in D$, but $fx = y$ does not.

Exercise

1. Show by an example that the following condition on a point y does not insure that $y \in fD$: if z is the center of D, then y and fz lie in a connected subset of $P - fC$.

16. Curves and closed curves

Heretofore the word "curve" has referred to the graph of a continuous function $f: [a, b] \to R$. We need now to use the word in the following broader sense. A *curve* in the plane is defined to be a mapping $\varphi: [a, b] \to P$ of some interval of real numbers into the plane. Each number $t \in [a, b]$ can be thought of as an instant of time, and the corresponding point $\varphi t \in P$ as the location of a moving point at the time t. Thus a curve may be regarded as the *path* of a moving point. In particular any curve has an orientation in the sense that the preferred or positive direction along the curve goes from φa to φb. This is the direction of motion (of increasing t). In pictures of the curves the orientation is indicated by arrowheads as in Fig. 16.1. Notice that we allow a curve to cross itself; that is, the moving point can pass through the same point at several different times. Moreover, the moving point may remain at rest for an interval of time. For example, the constant function which maps the entire interval $[a, b]$ into a single point is a curve in our sense. A *closed curve* is a curve which begins and ends at the same point: $\varphi a = \varphi b$.

The line segment L from a point A to a point B in P may be represented as a curve. Recall that any two segments are similar. So if $\varphi \colon [a, b] \to L$ is the similarity with $\varphi a = A$ and $\varphi b = B$, then φ defines a curve whose image is L. In this example the moving point has a constant velocity.

The graph of a continuous function $f \colon [a, b] \to R$ is a curve. We need only set φt equal to the point whose coordinates are (t, ft) for each $t \in [a, b]$. A curve of this type does not cross itself, nor is it closed, because $t_1 \neq t_2$ implies that φt_1 and φt_2 have different abscissas.

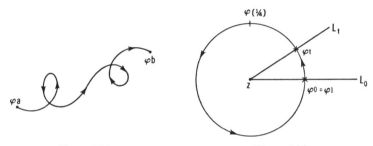

Figure 16.1 Figure 16.2

Any circle C is regarded as a closed curve in the following standard way. Let z denote the center of C, and let L_0 be a fixed ray (half-line including the initial point) issuing from z. Define φt for $t \in [0, 1]$ as follows: $\varphi 0$ is the point of intersection of L_0 with C, and φt is the point of C such that the angle at z between L_0 and the segment z to φt is $360t$ degrees. For example, $\varphi(\frac{1}{4})$ is the point at the 90° mark ($\frac{1}{4}$ the way around). (See Fig. 16.2.) Since there are 360° in a complete circle, we have $\varphi 1 = \varphi 0$. In this case also, the moving point has a constant speed.

The boundary of a rectangle may likewise be regarded as a closed curve. Take an interval $[a, e]$ and divide it into four subintervals by numbers b, c, d so that $a < b < c < d < e$. Let the four vertices of the rectangle be A, B, C, D in that order. As in the example above we can define φ so that it maps the intervals $[a, b]$, $[b, c]$, $[c, d]$ and $[d, e]$ onto the segments AB, BC, CD and DA, respectively. Since $\varphi a = A$ and $\varphi e = A$, the curve is closed.

Our pictorial illustrations foster the tendency, sometimes misleading, to regard a curve φ as being no more than the image $\varphi[a, b]$. So it must be emphasized that *the curve is the mapping* φ. For example, there is an infinity of distinct standard representations of the circle C as a closed curve, one for each choice of L_0.

Exercises

1. As a wheel of radius r with a flange rolls along a single rail without slipping, the center of the wheel traverses a straight line parallel to the track. Sketch the paths followed by

   (a) a point on the circumference;

   (b) a point at distance $r/2$ from the center;

   (c) a point at distance $5r/4$ from the center.

2. If $f: [a, b] \to P$ and $g: P \to P$, are continuous, show that gf is a curve.

3. If $f: [0, 1] \to [a, b]$ is the similarity such that $f0 = a$ and $f1 = b$, find the formula for ft when $t \in [0, 1]$. Give a formula for another such map which is not a similarity.

17. Intuitive definition of winding number

Let $\varphi: [a, b] \to P$ be a closed curve, let y be a point of the plane not on the curve, and, for each t in the interval, let L_t denote the ray starting at y and passing through φt. As t varies from a to b, the point φt traverses the curve, and the ray L_t rotates about its fixed endpoint at y. Since the curve is closed, L_t eventually returns to its initial position $L_b = L_a$. Therefore, during its motion, the ray made a whole number of complete rotations about y. This number is called the *winding number* of the closed curve φ about the point y, and we shall use the abbreviation $W(\varphi, y)$ for this number. By convention, counterclockwise rotations are given a positive sign and clockwise rotations a negative sign.

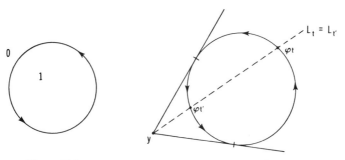

Figure 17.1   Figure 17.2

In Fig. 17.1, the circle, regarded as the closed curve described in Section 16, winds once about its center point. It also winds once about any point inside the circle. For any point outside the circle, the winding number is zero. As φt traces the circle, the ray L_t, pivoting on a point y outside the circle, sweeps over the portion of the plane bounded by the rays from y tangent to the circle. The sweeping goes first in one direction; then, after having reached one boundary (one of the tangent rays), reverses its direction, ultimately returning to its initial position without completing one rotation (see Fig. 17.2).

In Fig. 17.3, the closed curve is an ellipse traced once in the clockwise direction. For any point inside, the winding number is -1, and for any point outside, it is zero.

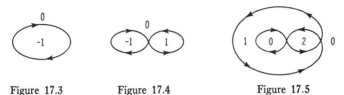

Figure 17.3   Figure 17.4   Figure 17.5

Fig. 17.4 is a closed curve in the shape of a figure 8 curve. The points in one bounded region all have winding number 1, and those in the other, -1. Of course all points in the unbounded region have winding number 0.

Fig. 17.5 shows the example discussed in Section 15. In this example, there is a bounded region about whose points the curve has winding number zero. For the other two bounded regions, the winding numbers are 1 and 2. Notice that, for two points in the same connected region, the winding numbers are always the same.

Fig. 17.6 shows the constant closed curve (all at a single point). It winds about each other point zero times.

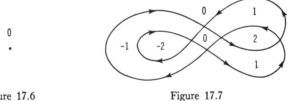

Figure 17.6   Figure 17.7

The diagram on the front cover of this book shows a closed curve and the winding numbers it has about points in each of its complementary regions.

Fig. 17.7 indicates that the possibilities are endless.

Now let C be a circle and let $f: C \to P$ be a mapping of C into the plane. Let $\varphi: [0, 1] \to C$ be the standard representation of C as a closed curve described in Section 16. Then the composition $f\varphi: [0, 1] \to P$ is again a closed curve because $\varphi 0 = \varphi 1$ implies that $f\varphi 0 = f\varphi 1$. It has a winding number about any point y of P not on fC. This is called the winding number of f about y and is denoted by $W(f, y)$.

Exercises

1. In Fig. 17.2, let y be at the distance $r\sqrt{2}$ from the center of the circle C of radius r. As we trace the outer arc of C from one point of tangency to the other, through what angle does the ray L_t turn? Through what angle does it turn as we continue around the circle along the inner arc?

2. The complement in P of the closed curve in Fig. 17.8 consists of seven connected regions labeled A, B, C, D, E, F, G. For each region, state the winding number of the closed curve about a point of that region.

3. Do the same as in the preceding problem for the closed curve of Fig. 17.9.

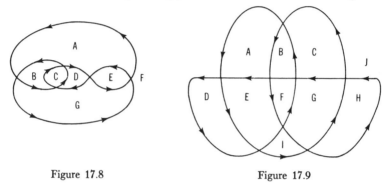

Figure 17.8   Figure 17.9

18. Statement of the main theorem

By using the concept of winding number, we can formulate now the main theorem of Part II.

THEOREM 18.1. *Let $f: D \to P$ be a mapping of a disk into the plane, let C be the boundary circle of D, and let y be a point of the plane not on fC. If the winding number of $f \mid C$ about y is not zero, then $y \in fD$; i.e. there is a point $x \in D$ such that $fx = y$.*

What follows is a short intuitive proof. Let r be the radius of C. For each number s such that $0 \leq s \leq r$, let C_s be the circle of radius s concentric with C; thus $C_r = C$, and C_0 is the center point z. Let y' be a point of the plane not in fD. Then for every s in $[0, r]$, y' is not on fC_s because C_s is in D, and so the winding number $W(f \mid C_s, y')$ is defined for every s in $[0, r]$. Abbreviate it by $W(s)$. Consider now the family of closed curves $f \mid C_s$ as s decreases from r to 0. Its members begin with $f \mid C$ and eventually shrink down to the constant curve $f \mid C_0$, i.e. to the point fz. Since $f \mid C_s$ varies gradually as s decreases steadily, it follows that $W(s)$ is a continuous function of $s \in [0, r]$.

How does the winding number $W(s)$ vary? The answer is: not at all, because W is a continuous function of s and each value $W(s)$ must be an integer; it cannot jump from one integer value to another without taking on non-integral values between (see the main theorem of Part I). Thus $W(s)$ has the same value for all s; in particular, $W(r) = W(0)$. But $W(0) = 0$ because $f \mid C_0 = fz$ is the constant closed curve. Therefore $f \mid C_r$ has winding number zero about y' for every point y' not in fD. It follows that $W(f \mid C_r, y) \neq 0$ implies that y is in fD; and to say that y is in fD is to say that there is an $x \in D$ such that $fx = y$.

One can see how the argument works in the illustration of Fig. 18.1 by following the successive closed curves $f \mid C_s$ as s decreases; as soon as the two lobes separate (e.g., the third closed curve drawn), y is clearly in the exterior of this closed curve, and therefore the winding number is zero. Notice that this agrees with the result obtained for Fig. 17.5.

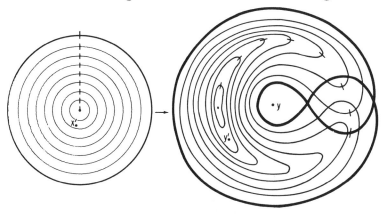

Figure 18.1

Exercises

1. If the closed curve in Exercise 2 of Section 17 is the $f \mid C$ of a map $f: D \to P$, which of the complementary regions $A, B, \cdots$ must lie in fD?

2. Answer the analogous question for the closed curve of Exercise 3, Section 17.

3. Let $f: P \to P$ be a mapping of the plane into itself that is a simple fold along a diameter of a disk.

    (a) What is the image of the boundary circle C? of the disk?

    (b) What is the winding number $W(f \mid C, y)$ of points y in the image of the disk?

19. When is an argument not a proof?

Most people, when they have seen and understood the arguments of the preceding two sections, are convinced that they have seen the truth and that little more need be added to achieve a complete and logical proof. However, a thoughtful reader should spot the gaps in the reasoning. The main gap occurs in Section 17; no precise definition of the winding number was given. It was left to the intuition to decide how many complete rotations the ray L_t makes about its endpoint y as t varies from a to b; it was assumed that our eyes could follow the rotating ray and integrate its motion into a single number of rotations. As is well known, eyesight is not entirely reliable in this respect; for example, we can be misled into thinking we are seeing continuous motion by a sufficiently rapid sequence of still pictures.

Fortunately mathematical concepts and deductions are independent of our ability to visualize motion. The situation we must treat is a static one. We have a closed curve φ, a point y not on the curve, and we wish to attach to y and φ an integer called the winding number which agrees with our intuitive notion. This will be done in the next seven sections. A reader, who prefers new ideas and applications to the careful development of an idea already outlined, should skip to Section 27.

Before we submerge ourselves in the details of the definition of $W(\varphi, y)$, let us note that, to complete the statement and proof of the main theorem, we need only

(1) define $W(\varphi, y)$ precisely,

(2) show that it is continuous under the type of variation of φ used in the intuitive proof presented in Section 18, and

(3) show that $W(\varphi, y) = 0$ whenever φ is a constant closed curve.

If we were to define $W(\varphi, y)$ to be 0 for all φ and y, this would satisfy the requirements (1), (2) and (3), and so the proof of the main theorem would be valid for this W, but the conclusion of the theorem would say nothing because there would be no points y such that $W(f \mid C, y) \neq 0$. Thus, to make our efforts worthwhile, we require also that

(4) $W(\varphi, y)$ should be non-zero for certain curves φ and points y; in particular, it should agree with the winding numbers defined intuitively in Section 17.

20. The angle swept out by a curve

In order to formulate a good definition of winding number, we first consider the more general concept of "angle swept out by a curve

§20] ANGLE SWEPT OUT BY CURVE

$\varphi\colon [a, b] \to P$ with respect to a point y". We shall define the measure $A(\varphi, y)$ of this angle in two stages: first for a special class of curves, then for any curve. (Unless otherwise stated, angles will be measured in degrees, and the degree symbol ° will be omitted.) We shall then see that the measure of an angle swept out by a closed curve is a multiple of 360, and this multiple will be defined to be the winding number $W(\varphi, y)$:

$$W(\varphi, y) = \frac{A(\varphi, y)}{360}.$$

The special class of curves for which $A(\varphi, y)$ can be defined readily and unambiguously consists of the so-called short curves relative to the point y not on the curve; φ is called *short relative to* y if there is a ray L issuing from y which does not meet the curve. For example, a constant curve $\varphi t = z \neq y$ for all $t \in [a, b]$ is short relative to y. Fig. 20.1 shows a less trivial example of a short curve. As t varies from a to b, φt varies from φa to φb along the curve, and the ray issuing from y to the point φt on the curve rotates from L_a to L_b. Let $\measuredangle L_a L_b$ denote this angle of rotation; since φ is short, this angle does not include the ray L (from y and not meeting φ). We define $A(\varphi, y)$ to be the measure, in degrees, of $\measuredangle L_a L_b$; a counterclockwise angle is given a positive sign, and a clockwise angle a negative sign.

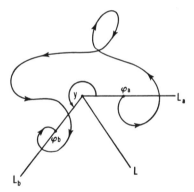

Figure 20.1

Suppose we have a protractor in the form of a complete circle C divided into 360 equal arcs, the points of division being numbered from a zero-point in counterclockwise order from 0 to 359. We place C so that its center is at y and it is zeroed along the ray L (i.e. the zero-point on C is at the intersection of C with L). Let p_a, p_b denote the intersections of C with the rays L_a, L_b, respectively, and let x_a, x_b denote their respective protractor readings; then the degrees in the angle $L_a L_b$ can be computed by

$$A(\varphi, y) = x_b - x_a.$$

Observe that the difference $x_b - x_a$ is independent of the position of the initial ray of the protractor provided that this ray is not contained in the angle A we are measuring (see Fig. 20.2); for, suppose we rotate our protractor by an angle α so that it is now zeroed at a ray L'. The new protractor readings are

$$x'_a = x_a - \alpha \quad \text{and} \quad x'_b = x_b - \alpha,$$

so that

$$x'_b - x'_a = x_b - x_a = A(\varphi, y)$$

provided L, L' do not cut the arc $p_a p_b$ of C carrying the radial projection of the curve. It is easy to see that, under these conditions, the protractor readings are always numbers between 0 and 360, and that the difference $x_b - x_a$ is always between -360 and 360. In Fig. 20.1, it is positive and about 230. (If the orientation of the curve were reversed, i.e., if φa and φb were interchanged, then $A(\varphi, y)$ would be about -230.)

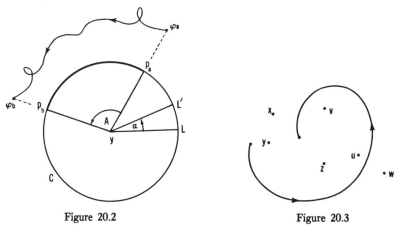

Figure 20.2   Figure 20.3

The formula

$$A(\varphi, y) = x_b - x_a$$

thus gives the measure of the angle swept out by a short curve φ with respect to a point y uniquely. Moreover, the definition of "short" curve guarantees that there is a ray L from y not meeting φ along which we may zero our protractor to compute $A(\varphi, y)$; if there are many such rays, it does not matter which one we select as initial ray of our protractor.

Notice how our definition provides for the cancellation of positive and negative motions. For example, in Fig. 20.1, the line L_t at the start rotates through an angle of $-30°$ and immediately cancels this by rotating back to its initial position. Similarly, as the point traverses the loop at the top of the curve, the rotation of L_t adds up to zero.

In case φ is a constant curve mapping $[a, b]$ into a single point, then φ is a short curve and $A(\varphi, y) = 0$ because $L_a = L_b$ and so $x_a = x_b$.

The function $A(\varphi, y)$ has a most important property called *additivity* in φ. Let $a < b < c$ be real numbers, and let $\varphi: [a, c] \to P$ be a short curve relative to y. Let φ_1, φ_2 be the two parts of φ obtained by restricting φ to $[a, b]$ and to $[b, c]$ respectively. We can think of φ as the union of φ_1 and φ_2. Clearly φ_1, φ_2 are also short relative to y. Choosing a ray L issuing from y and missing φ, we obtain protractor readings x_a, x_b, x_c for L_a, L_b, L_c, respectively. Since

$$x_c - x_a = (x_b - x_a) + (x_c - x_b),$$

we see that

$$A(\varphi, y) = A(\varphi_1, y) + A(\varphi_2, y).$$

Exercises

1. In Fig. 20.3, for which of the points u, v, w, x, y, z is the curve C short?

2. For each point for which C is short, what is the angle in degrees?

21. Partitioning a curve into short curves

If $\varphi: [a, b] \to P$ is a curve, we can cut it up into pieces by dividing $[a, b]$ into subintervals and restricting φ to each of these in turn. In this way φ can be decomposed into a union of shorter curves. In case φ is not short relative to a point y, it may happen that each of the pieces of such a decomposition is short relative to y. Then, by adding the measures of the angles subtended at y by the various pieces, we can obtain a value for $A(\varphi, y)$.

To be precise, a decomposition of $\varphi: [a, b] \to P$ into a union of curves is called a *partition* $\mathcal{P}$ of φ. It consists first of an increasing sequence of numbers starting with a and ending with b

$$a = t_0 < t_1 < \cdots < t_{m-1} < t_m = b,$$

and secondly of the sequence of curves $\varphi_1, \varphi_2, \cdots, \varphi_m$ where φ_i denotes the restriction of φ to the interval $[t_{i-1}, t_i]$ ($i = 1, 2, \cdots, m$). The partition is called *sufficiently fine* for a point y not on φ if each of the pieces φ_i is short relative to y. In that case, each of the $A(\varphi_i, y)$ is defined and their sum is denoted by $A(\mathcal{P}, y)$:

(21.1)
$$A(\mathcal{P}, y) = \sum_{i=1}^{m} A(\varphi_i, y)$$
$$= A(\varphi_1, y) + A(\varphi_2, y) + \cdots + A(\varphi_m, y).$$

We shall prove two propositions:

1. If φ is any curve and y a point not on φ, then there is a partition sufficiently fine for y.

2. If $\mathcal{P}$ and $\mathcal{P}'$ are any two partitions of φ sufficiently fine for y, then $A(\mathcal{P}, y) = A(\mathcal{P}', y)$.

Once these facts are proved we can define $A(\varphi, y)$ for any curve φ as follows:

DEFINITION. If φ is a curve in the plane and y a point not on φ, the common value of $A(\mathcal{P}, y)$ for all sufficiently fine partitions $\mathcal{P}$ of φ is the measure of *the angle at y swept out by φ*. It is denoted by $A(\varphi, y)$ and may be computed from the formula (21.1), each term of the sum on the right being computed by the method of Section 20.

The first proposition tells us that we can find a partition for which $A(\mathcal{P}, y)$ is defined. The second tells us that the number $A(\mathcal{P}, y)$ so obtained doesn't depend on which partition we may have chosen, and so it depends only on φ and y.

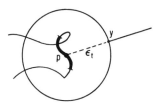

Figure 21.1

PROOF OF 1. At any point $p = \varphi t$ of the curve, the circle with center p and radius $d(p, y)$ passes through v (see Fig. 21.1). Any piece of the curve lying inside this circle is short relative to y because it does not meet the ray issuing from y which is the continuation of the segment py leading out of the circle. Let us apply the continuity of φ at t taking $\epsilon_t = d(p, y)$. It provides us with a number $\delta_t > 0$ such that $\varphi_{t'} \in N(p, \epsilon_t)$ for every $t' \in N(t, \delta_t)$. It follows now that, for every interval $I' \subset N(t, \delta_t)$, the curve $\varphi \mid I'$ is short relative to y.

Since $[a, b]$ is compact and the neighborhoods $\{N(t, \delta_t)\}$ cover $[a, b]$, there is a finite number of these neighborhoods, say $N_1, N_2, \cdots, N_k$, covering $[a, b]$. Let S be the set of all endpoints of the open intervals $N_1, N_2, \cdots, N_k$. Let δ be half the shortest of the distances $d(s, t)$ for $s, t \in S$ and $s \neq t$. Let $\mathcal{P}$ be any partition of $[a, b]$ by intervals of lengths at most δ. We claim that $\mathcal{P}$ is sufficiently fine for y. To prove this, it suffices to show that every subinterval I' of $\mathcal{P}$ is contained in some one of the $N_1, N_2, \cdots, N_k$. Since the length of I' is at most δ, I' contains either no point of S or one point of S, say c. In the first case, choose any N_i which meets I' (I' is covered by $N_1, \cdots, N_k$); then $I' \subset N_i$ because the interval I' contains neither endpoint of the open interval N_i. In the second case, choose any N_i containing c; then again $I' \subset N_i$ because I' contains neither endpoint of N_i. This completes the proof of 1.

PROOF OF 2. If $\mathcal{P}$ is a sufficiently fine partition of φ, and $\mathcal{P}'$ is a partition obtained from $\mathcal{P}$ by introducing a new point into a subinterval, say I_k, dividing it into two subintervals I', I'', then the additivity proved in Section 20 tells us that

$$A(\varphi \mid I', y) + A(\varphi \mid I'', y) = A(\varphi \mid I_k, y).$$

Adding to both sides the terms $A(\varphi \mid I_j, y)$ for all $j \neq k$ leads to the result $A(\mathcal{P}, y) = A(\mathcal{P}', y)$.

Let us call the points $\{t_0, t_1, \cdots, t_m\}$ of a partition $\mathcal{P}$ the *vertices* of $\mathcal{P}$. If $\mathcal{P}, \mathcal{P}'$ are two partitions of φ such that the vertices of $\mathcal{P}$ are contained in those of $\mathcal{P}'$, then $\mathcal{P}'$ is called a refinement of $\mathcal{P}$, and we express this by writing $\mathcal{P}' < \mathcal{P}$. Each subinterval of $\mathcal{P}'$ must be included in some subinterval of $\mathcal{P}$. So if $\mathcal{P}$ is sufficiently fine so are all refinements of $\mathcal{P}$.

If $\mathcal{P}' < \mathcal{P}$, we may take the vertices of $\mathcal{P}'$ which are not vertices of $\mathcal{P}$ and adjoin them one at a time. We obtain thus a sequence of refinements $\mathcal{P} = \mathcal{P}_1 > \mathcal{P}_2 > \cdots > \mathcal{P}_s = \mathcal{P}'$. Assume moreover that $\mathcal{P}$ is sufficiently fine. Then the result of the first paragraph of this proof gives

$$A(\mathcal{P}, y) = A(\mathcal{P}_1, y) = A(\mathcal{P}_2, y) = \cdots = A(\mathcal{P}_s, y) = A(\mathcal{P}', y).$$

Therefore $\mathcal{P}' < \mathcal{P}$ and $\mathcal{P}$ sufficiently fine imply $A(\mathcal{P}, y) = A(\mathcal{P}', y)$.

Now let $\mathcal{P}_1, \mathcal{P}_2$ be any two sufficiently fine partitions. The union of the vertices of $\mathcal{P}_1$ and of $\mathcal{P}_2$ is the set of vertices of a partition that we shall call $\mathcal{P}_3$. Clearly $\mathcal{P}_3 < \mathcal{P}_1$ and $\mathcal{P}_3 < \mathcal{P}_2$. The result of the preceding paragraph gives $A(\mathcal{P}_1, y) = A(\mathcal{P}_3, y)$ and $A(\mathcal{P}_2, y) = A(\mathcal{P}_3, y)$; hence $A(\mathcal{P}_1, y) = A(\mathcal{P}_2, y)$, and our proof is complete.

Exercises

1. For a point y in each of the regions A, B, D, and F of Fig. 17.8, find a partition which is sufficiently fine for y.

2. Let the curve in Fig. 21.2 be partitioned by the points a, b, c, d, e, f as shown.

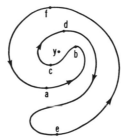

Figure 21.2

(a) Beginning at a and following the arrows, what is the largest section of the curve that is short relative to y?

(b) Is there another section that may be adjoined to the section from a to d for which the curve is short relative to y?

(c) Find the least number of the points $a, b, \cdots$ which provide a partition sufficiently fine for y.

22. The winding number $W(\varphi, y)$

Once a sufficiently fine partition $\mathcal{P} = \{\varphi_0, \varphi_1, \cdots, \varphi_m\}$ of φ relative to y has been found, the computation of $A(\varphi, y)$ is a routine affair of computing each $A(\varphi_i, y)$ with the protractor and adding the results with due regard for signs. In measuring $A(\varphi_i, y)$, the zero of the scale is set on a ray L_i which misses φ_i, and the difference is taken of the two readings at the endpoints of φ_i. Thus the protractor must be repositioned for each short curve. We shall show now how all the readings can be taken with a single position of the protractor, and how the computation can be greatly shortened.

As before, let C be the circle with center y and radius 1. For each $i = 0, 1, \cdots, m$, let p_i denote the point of intersection of C with the ray from y through the point φt_i of the curve φ. For each short curve φ_i, let L_i be a ray from y which does not intersect φ_i. Two points on the circle, p_{i-1} and p_i, determine two arcs; of these two, let $p_{i-1}p_i$ denote the arc not intersecting L_i. Then, as shown in Section 20, $A(\varphi_i, y)$ is the angular measure of this arc with due regard for sign. The arc is oriented from p_{i-1} to p_i. If this orientation is counterclockwise, the sign of $A(\varphi_i, y)$ is positive, otherwise it is negative.

Assume for convenience that our protractor has radius 1. Let us center it at y and rotate it until the zero is at a point q of C different from $p_0, p_1, \cdots, p_m$. Now keep the protractor fixed, and let $x_0, x_1, \cdots, x_m$

denote the direct readings in degrees for the points $p_0, p_1, \cdots, p_m$, respectively. Each x_i lies between 0 and 360. We can now state our simplified formula for $A(\varphi, y)$.

THEOREM 22.1. *Let r be the number of arcs $p_{i-1}p_i$ which contain q and have positive orientation, and let s be the number of arcs $p_{i-1}p_i$ which contain q and have negative orientation. Then*

$$A(\varphi, y) = x_m - x_0 + (r - s)360.$$

An illustration of the theorem is given in Fig. 22.1. If we draw the ray yq, we see that it meets the curve φ three times. At each point of intersection, the orientation of φ, and hence of the arc $p_{i-1}p_i$, can be determined. It is first negative, then negative again, and finally positive; so $r = 1$ and $s = 2$. The protractor readings, zeroed at q, are 65 for x_m and 195 for x_0. So $A(\varphi, y) = 65 - 195 + (1 - 2)360 = -490$.

It is a consequence of the theorem that the number

$$x_m - x_0 + (r - s)360$$

does not depend on the choice of the zero-point q even though x_m, x_0, r and s do depend on that choice.

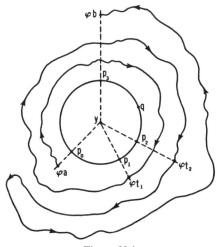

Figure 22.1

To prove the theorem, recall that

$$A(\varphi, y) = \sum_{i=1}^{m} A(\varphi_i, y).$$

We shall express each $A(\varphi_i, y)$, which is the angular measure of the arc $p_{i-1}p_i$, in terms of the readings x_{i-1}, x_i.

Consider first an i such that the arc $p_{i-1}p_i$ does not contain q as in

Fig. 22.2. By our definition of the measure of the angle swept out by a short curve, $A(\varphi, y) = x_i - x_{i-1}$. Note that this holds even when the arc has a negative orientation, for then $x_{i-1} > x_i$, and so $x_i - x_{i-1}$ is negative.

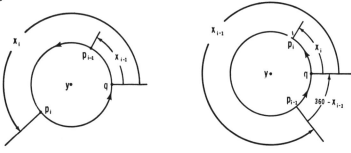

Figure 22.2   Figure 22.3

Consider next an i such that the arc $p_{i-1}p_i$ is positively oriented and contains q as in Fig. 22.3. By adding the angles determined by the arcs $p_{i-1}q$ and qp_i, we obtain

$$A(\varphi_i, y) = 360 - x_{i-1} + x_i = x_i - x_{i-1} + 360.$$

Consider finally an i such that the arc $p_{i-1}p_i$ is negatively oriented and contains q as in Fig. 22.4. By adding the angles determined by the arcs $p_{i-1}q$ and qp_i, we obtain

$$A(\varphi_i, y) = -x_{i-1} - (360 - x_i) = x_i - x_{i-1} - 360.$$

The three cases above comprise all possibilities. If we add all the $A(\varphi_i, y)$, $i = 1, 2, \cdots, m$, each term will have an $x_i - x_{i-1}$, r of the terms will have a $+360$, and s of the terms a -360. Therefore

$$A(\varphi, y) = (x_m - x_{m-1}) + (x_{m-1} - x_{m-2}) + \cdots$$
$$+ (x_1 - x_0) + r360 - s360$$
$$= x_m - x_0 + (r - s)360,$$

and this completes the proof.

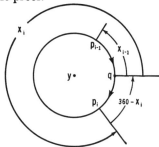

Figure 22.4

COROLLARY. *If φ is a closed curve, then $A(\varphi, y) = (r - s)360$.*

The proof of the corollary lies merely in noting that $x_m = x_0$ for a closed curve because $\varphi a = \varphi b$.

We are finally in a position to give a precise definition of winding number: $W(\varphi, y) = A(\varphi, y)/360 = r - s$. It follows that the winding number is an integer.

Exercises

1. If a closed curve is short relative to a point y, what is its winding number?

2. The closed curve in Fig. 22.5 has been partitioned by highest and lowest points as indicated, and the protractor readings for the vertices are given in the following table, with the protractor zeroed along the indicated ray.

$$\varphi t_i = a, \quad b, \quad c, \quad d, \quad e, \quad f, \quad g, \quad h$$
$$x_i = 270, \; 90, \; 300, \; 20, \; 340, \; 45, \; 350, \; 60$$

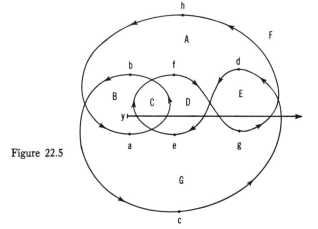

Figure 22.5

(a) Find the angle at y swept out by the curve from a to d; from b to g.

(b) Find the winding number $W(\varphi, y)$ and verify the result obtained in Exercise 2, Section 17.

(c) Suggest a different direction for the ray yq that will reduce the computation of $W(\varphi, y)$ to a minimum.

(d) For a point in each of the regions $A, B, \cdots$, verify the results for the winding number obtained in Section 17.

23. Properties of $A(\varphi, y)$ and $W(\varphi, y)$

Having defined $A(\varphi, y)$ and $W(\varphi, y)$ precisely, we must now show that they have the properties we claimed for them.

1. If φ is a constant curve, then $A(\varphi, y) = 0$ and $W(\varphi, y) = 0$.

Since $\varphi[a, b]$ is a point, φ is a short curve, and the trivial partition is sufficiently fine. But for short curves $\varphi a = \varphi b$ implies $x_a = x_b$, so $A(\varphi, y) = 0$; hence $W(\varphi, y) = A(\varphi, y)/360 = 0$.

2. $A(\varphi, y)$ is additive in φ. Precisely, suppose $a < b < c$, and $\varphi: [a, c] \to P$. Let $\varphi_1 = \varphi \,|\, [a, b]$, and $\varphi_2 = \varphi \,|\, [b, c]$. Then $A(\varphi, y) = A(\varphi_1, y) + A(\varphi_2, y)$. In case $\varphi a = \varphi b = \varphi c$, $\varphi_1, \varphi_2, \varphi$ are closed curves, and $W(\varphi, y) = W(\varphi_1, y) + W(\varphi_2, y)$.

Let $\mathcal{P}_1$ and $\mathcal{P}_2$ be sufficiently fine partitions of φ_1, φ_2 respectively. Then the union of the vertices of $\mathcal{P}_1, \mathcal{P}_2$ gives a sufficiently fine partition $\mathcal{P}$ of φ. Since the terms of the sum $A(\mathcal{P}, y)$ are the same as those of the sum $A(\mathcal{P}_1, y) + A(\mathcal{P}_2, y)$ it is evident that

$$A(\mathcal{P}, y) = A(\mathcal{P}_1, y) + A(\mathcal{P}_2, y),$$

and this proves the first relation. In case $\varphi, \varphi_1, \varphi_2$ are closed curves, each term in this relation is an integer multiple of 360. Dividing by 360 gives the additivity of the winding number in φ.

Exercise

1. Use the figure and protractor readings given in Exercise 2, Section 22. If $\varphi t_0 = a$, $\varphi t_1 = d$, and $\varphi t_2 = g$, find $A(\varphi_1, y)$, $A(\varphi_2, y)$, and apply the appropriate result of this section to find $A(\varphi \,|\, [t_0, t_2], y)$.

24. Homotopies of curves

In the next section we shall show that the winding number of a curve about a point does not change if the curve or the point is varied in a continuous fashion (see Section 18). Our purpose in this section is to describe precisely the kind of variation we shall allow.

DEFINITION. Let φ_0 and φ_1 be two curves in a space Y which are defined on the same interval $[a, b]$. Then a homotopy of φ_0 into φ_1 is a mapping Φ of a rectangle Q into Y so that the lower edge of Q maps onto the curve φ_0, and the upper edge onto φ_1. Precisely, let Q be the rectangle in the plane of two variables (t, τ) such that $a \leq t \leq b$ and $0 \leq \tau \leq 1$. Then a homotopy Φ of φ_0 into φ_1 is a mapping $\Phi\colon Q \to Y$ such that

$$\Phi(t, 0) = \varphi_0 t \quad \text{and} \quad \Phi(t, 1) = \varphi_1 t \quad \text{for all } t \in [a, b].$$

In case φ_0, φ_1 are closed curves, a homotopy of φ_0 into φ_1 as *closed* curves is a homotopy Φ, as above, satisfying the additional condition

$$\Phi(a, \tau) = \Phi(b, \tau) \quad \text{for all } \tau \in [0, 1].$$

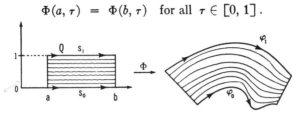

Figure 24.1

To make this definition transparent, picture the rectangle Q as composed of horizontal line segments s_τ, one for each value $\tau \in [0, 1]$. The restriction of Φ to one of the segments s_τ determines a curve $\varphi_\tau\colon [a, b] \to Y$ defined by $\varphi_\tau t = \Phi(t, \tau)$. We obtain thus a family of curves, one for each value of τ between 0 and 1 (see Fig. 24.1). If we regard τ as the time variable, we can think of the family of curves as the various positions of a single moving curve. Each vertical segment in Q maps into the path followed by a point of the moving curve. Because of this picture, a homotopy is often called a *deformation*.

If the two curves φ_0, φ_1 map $[a, b]$ into the plane P or into R^m, then there is a special homotopy of φ_0 into φ_1 called the *linear* homotopy. For each pair (t, τ) in Q, define $\Phi(t, \tau)$ to be the point which divides the line segment from $\varphi_0 t$ to $\varphi_1 t$ in the ratio $\tau : 1 - \tau$ (see Fig. 24.2). The ratio 0:1 gives the initial end of the segment, and 1:0 the terminal; hence $\Phi(t, 0) = \varphi_0 t$ and $\Phi(t, 1) = \varphi_1 t$. The restriction of Φ to each vertical segment of Q is a similarity because preservation of ratios is characteristic of a similarity.

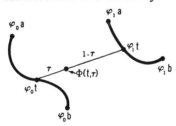

Figure 24.2

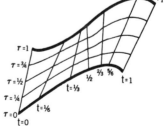

Figure 24.

An illustration of a linear homotopy is given in Fig. 24.3. The positions of the moving curve are drawn for the times $\tau = 0, \frac{1}{4}, \frac{1}{2}, \frac{3}{4}$, and 1; and the straight line paths followed by various points on the curve are shown for $t = 0, \frac{1}{6}, \frac{1}{3}, \frac{1}{2}, \frac{2}{3}, \frac{5}{6}$, and 1. Notice that a single point of the moving curve follows a straight line at a constant velocity.

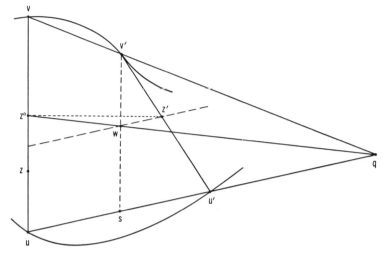

Figure 24.4

We must prove that the linear homotopy Φ is continuous.† Let (c, γ) denote the coordinates of a point of Q at which we wish to prove continuity. Let (t, τ) denote the coordinates of any other point of Q. Introduce the abbreviations:

$$u = \varphi_0 c, \quad v = \varphi_1 c, \quad z = \Phi(c, \gamma), \quad z'' = \Phi(c, \tau),$$
$$u' = \varphi_0 t, \quad v' = \varphi_1 t, \quad z' = \Phi(t, \tau).$$

Fig. 24.4 gives a picture of the situation when (t, τ) is near to (c, γ). The solid line segments are the paths followed under the homotopy by the points φc and φt as τ varies from 0 to 1. We wish to show that the distance $d(z, z')$ can be made small (less than a prescribed $\epsilon > 0$) by restricting (t, τ) to be near (c, γ). By the triangle inequality,

$$d(z, z') \leq d(z, z'') + d(z'', z').$$

Since the mapping of $[0, 1]$ into the segment uv which sends τ into $\Phi(c, \tau)$ is a similarity, it is continuous. Hence, corresponding to the positive number $\epsilon/2$, there is a $\delta' > 0$ such that

---

† For a reader who is familiar with vector algebra, we can write $\Phi(t, \tau) = (1 - \tau)(\varphi_0 t) + \tau(\varphi_1 t)$, and we can argue that $(1 - \tau)(\varphi_0 t)$ is continuous because it is the product of the continuous scalar function $1 - \tau$ and the continuous vector function $\varphi_0 t$. Similarly $\tau(\varphi_1 t)$ is continuous. Finally, $\Phi(t, \tau)$ is continuous because it is the vector sum of two continuous functions.

$d(z, z'') < \epsilon/2$ for every τ satisfying $|\tau - \gamma| < \delta'$.

Since φ_0, φ_1 are continuous at c, there are numbers $\delta_0 > 0$ and $\delta_1 > 0$ such that

$d(u, u') < \epsilon/2$ for every t satisfying $|t - c| < \delta_0$,

$d(v, v') < \epsilon/2$ for every t satisfying $|t - c| < \delta_1$.

Now let δ be the smallest of δ', δ_0 and δ_1. If (t, τ) is in $N((c, \gamma), \delta)$, then all three of the preceding inequalities hold. It can be shown that $d(z'', z')$ is no greater than the larger of $d(u, u')$ and $d(v, v')$. Fig. 24.4 shows how this is proved when $d(u, u')$ is the larger. Through v' construct a parallel to vu meeting uu' in s, and through z' construct a parallel to uu' meeting $v's$ in w. Then line qw, extended, meets vu in a point r which, by similar triangles, divides vu in the same ratio as z' divides $v'u'$; hence $r = z''$. Then

$$d(z'', z') \leq d(z'', w) + d(w, z')$$

$$\leq d(u, s) + d(s, u') = d(u, u').$$

Therefore $d(z'', z') < \epsilon/2$. Combining this with $d(z, z'') < \epsilon/2$ and the triangle inequality gives $d(z, z') < \epsilon$. This completes the proof of the continuity of Φ.

In case the two curves φ_0, φ_1 are closed, the linear homotopy of φ_0 into φ_1 is a homotopy as closed curves; for, $\varphi_0 a = \varphi_0 b$ and $\varphi_1 a = \varphi_1 b$ imply that the segments $\varphi_0 a$ to $\varphi_1 a$ and $\varphi_0 b$ to $\varphi_1 b$ coincide, and so $\Phi(a, \tau) = \Phi(b, \tau)$ for all $\tau \in [0, 1]$.

Whenever the end curve φ_1 of a homotopy is a constant curve at a point, then the homotopy is said to *shrink* the initial curve φ_0 to a point. An important example of this is the following. Let D be a disk with center z and boundary circle C. Let $\varphi_0: [0, 1] \to C$ be the standard representation of C as a closed curve; that is, $[0, 1]$ is wrapped just once around C in the counterclockwise sense (see Section 16). Let $\varphi_1: [0, 1] \to z$ be the constant curve at the center. Finally let Φ be the linear homotopy of φ_0 into φ_1. Then Φ shrinks φ_0 (or C) to a point. If we picture the homotopy as a moving curve, then at each instant, it is a circle with center z, and each point moves towards z along a radial line.

Now let $f: D \to P$ be a mapping of the disk into the plane, and let Φ be the homotopy just described. Then the composition $f\Phi: Q \to P$ is a shrinking of the closed curve $f\varphi_0: [a, b] \to P$ into the constant closed curve $f\varphi_1$ at fz. This homotopy gives the family of closed curves used in the intuitive proof of our main theorem (Section 18).

Exercises

1. Show that any closed curve in the plane is homotopic to a constant closed curve.

2. Show that a curve $\varphi: [a, b] \to Y$ in any space Y is homotopic to a constant mapping leaving one endpoint fixed.

3. Let $a < b < c \in R$, let $\varphi: [a, c] \to Y$ be such that $\varphi a = \varphi b = \varphi c$, and suppose the closed curves $\varphi \mid [a, b]$, $\varphi \mid [b, c]$ are homotopic to constants leaving fixed endpoints; then φ is homotopic to a constant mapping leaving ends fixed.

4. Show that a homotopy Φ of φ_0 into φ_1 can be reversed to give a homotopy of φ_1 into φ_0.

25. Constancy of the winding number

THEOREM 25.1. *Let $\Phi: Q \to P$ be a homotopy of φ_0 into φ_1 as closed curves, and let y be a point not in the image ΦQ. Then the winding number $W(\varphi_\tau, y)$ is constant as τ varies from 0 to 1. In particular, $W(\varphi_0, y) = W(\varphi_1, y)$.*

Let us abbreviate $W(\varphi_\tau, y)$ by $f\tau$; f is a function defined on $[0, 1]$, and, as shown in Section 22, each value of f is an integer. The main part of our proof consists in showing that f is constant under "small changes" of τ. Precisely, if $\alpha \in [0, 1]$, then there is a neighborhood N_α of α such that $f\tau = f\alpha$ for every $\tau \in N_\alpha$. Once this has been done the theorem is proved as follows: Since f is constant on N_α, it is continuous on N_α and hence continuous at α; and since this is true for each α, f is continuous in $[0, 1]$. Now, if f were not constant and had at least two different values, then the main theorem of Part I would say that f takes on all values between the two values, including the non-integral numbers between. This contradicts the fact that every value of f is an integer, so it follows that f must be constant.

To prove constancy near $\alpha \in [0, 1]$, choose a partition $\mathcal{P}$ sufficiently fine for φ_α. First we shall find a neighborhood N' of α such that, for all $\tau \in N'$, $\mathcal{P}$ is sufficiently fine for φ_τ. Then, using the method of computing winding numbers described in Section 22, we shall find a still smaller neighborhood N such that *each separate step of the computation remains constant* in N.

Let $t_0, t_1, \cdots, t_m$ be the vertices of $\mathcal{P}$. For each subinterval $I_k = [t_{k-1}, t_k]$, the fineness of $\mathcal{P}$ means that there is a ray L_k from y not meeting $\varphi_\alpha I_k$. Let D_k denote the line segment in Q from (t_{k-1}, α)

to (t_k, α). We shall show that there is a rectangle E_k containing D_k such that its image ΦE_k does not meet the ray L_k. Let

$$V_k = \Phi^{-1}(P - L_k).$$

Since Φ is continuous and $P - L_k$ is an open set, it follows that V_k is open. Since $\Phi D_k = \varphi_\alpha I_k$ is in $P - L_k$, it follows that $D_k \subset V_k$. For each point $p \in D_k$, we may choose a circular neighborhood $N(p) \subset V_k$. Then let $M(p)$ denote the interior of the largest square in $N(p)$ whose sides are parallel to the (t, τ)-axes. The collection of these $M(p)$ for all $p \in D_k$ is an open covering of D_k. Since D_k is compact, this covering contains a finite covering, say $M_1, M_2, \cdots M_n$. Let δ_k denote half the width of the smallest of these squares, and let E_k denote the rectangle of (t, τ)-values such that $t \in I_k$ and

$$|\tau - \alpha| < \delta_k.$$

By our choice of δ_k, it follows that

$$E_k \subset M_1 \cup M_2 \cup \cdots \cup M_n \subset V_k.$$

This means that $\varphi_\tau I_k \subset P - L_k$ for every $\tau \in N(\alpha, \delta_k)$. Supposing this construction done for each of the intervals I_k, $k = 1, 2, \cdots, m$, let δ be the least of the numbers $\delta_1, \delta_2, \cdots, \delta_m$, and let $N' = N(\alpha, \delta)$. Then, if $\tau \in N'$ we have that $\varphi_\tau I_k \subset P - L_k$ for all $k = 1, 2, \cdots, m$. This proves that, for every $\tau \in N'$, $\mathcal{P}$ is sufficiently fine for φ_τ.

We now apply the method of Section 22 for computing winding numbers to this partition $\mathcal{P}$ and the various curves φ_τ for $\tau \in N'$. As in Section 22, C denotes the circle with center y and radius 1. Let $g: P - y \to C$ denote the radial projection from y onto C. Then $g\Phi: Q \to C$ and $g\varphi_\tau: [a, b] \to C$ are continuous because they are compositions of continuous functions. The image on C of the vertex t_k of $\mathcal{P}$ under $g\varphi_\tau$, namely $g\varphi_\tau t_k$, is abbreviated by $p_k\tau$. Let $A_k\tau$ denote the arc of C from $p_{k-1}\tau$ to $p_k\tau$ which does not meet L_k. If q is a point of C distinct from $p_0\tau, p_1\tau, \cdots, p_m\tau$, then, as in Section 22, the winding number $W(\varphi_\tau, y)$ is $f\tau = r - s$, where r (respectively s) is the number of positively (negatively) oriented arcs $A_k\tau$ which contain q.

Select a $q \in C$ distinct from $p_0\alpha, p_1\alpha, \cdots, p_m\alpha$. For each $k = 1, 2, \cdots, m$, choose a neighborhood U_k of $p_k\alpha$ on C which does not contain q and does not meet either L_k or L_{k+1}. Then U_k will be a short arc of C containing $p_k\alpha$. Since $g\Phi$ is continuous and U_k is a neighborhood of $p_k\alpha$, and $g\Phi(t_k, \alpha) = p_k\alpha$, it follows that there is a neighborhood N_k of α in $[0, 1]$ such that $g\Phi(t_k, \tau) = p_k\tau$ lies in U_k for all $\tau \in N_k$. Let N denote the smallest of the neighborhoods N' and $N_1, N_2, \cdots, N_m$. Then $\tau \in N$ implies that $p_k\tau \in U_k$ for every $k = 1, 2, \cdots, m$. (Note that $p_m\tau = p_0\tau$ because each φ_τ is a closed curve.) In particular, q is distinct from $p_0\tau, p_1\tau, \cdots, p_{m-1}\tau$ because q is in none of the U_k's.

It remains to show that, for each $\tau \in N$ and each $k = 1, 2, \cdots, m$, the two arcs $A_k\tau$ and $A_k\alpha$ bear the same relation to q; for then we obtain the same count $r - s$ for φ_τ as for φ_α. Consider first the case of a k where U_{k-1} and U_k have a point in common. Then $U_{k-1} \cup U_k$ is a connected arc, and its complement D in C is a connected arc which meets L_k, contains q, but for all $\tau \in N$, contains neither $p_{k-1}\tau$ nor $p_k\tau$. Since D is connected, it lies wholly in one of the two arcs of C from $p_{k-1}\tau$ to $p_k\tau$; one of these is $A_k\tau$, and the other meets L_k. Since D meets L_k, it follows that D does not meet $A_k\tau$. Since q is in D, this proves that, for all $\tau \in N$, $A_k\tau$ does not contain q. Thus, in this case, the relationship of $A_k\tau$ to q is constant for $\tau \subset N$; that is, if $A_k\alpha$ does not contain q, neither does $A_k\tau$ for $\tau \in N(\alpha)$.

Consider finally a k such that U_{k-1} and U_k have no common point. Then the complement of $U_{k-1} \cup U_k$ in C consists of two arcs D and E, and we let D denote the one which meets L_k. Arguing as above, we see that, for all $\tau \in N$, D does not meet $A_k\tau$. Moreover, since E does not meet L_k, all of the arcs $A_k\tau$, for $\tau \in N$, contain E and are similarly oriented from U_{k-1} to U_k. Now q is not in $U_{k-1} \cup U_k$ so it must lie in $D \cup E$. If $q \in D$, none of the arcs $A_k\tau$, for $\tau \in N$, contain q. If $q \in E$, all of the arcs $A_k\tau$, for $\tau \in N$, contain q and are similarly oriented. Thus the relationship of $A_k\tau$ to q is constant as τ varies in N. This completes the proof of the constancy for all $\tau \in N$ of the winding number $f\tau$. By the argument immediately following Theorem 25.1, this proves the constancy of the winding number under a homotopy of the curve.

THEOREM 25.2. *Let $\varphi: [a, b] \to P$ be a closed curve in the plane. Let y_0, y_1 be two points of P not on the curve φ such that y_0, y_1 can be joined by a curve $\psi: [0, 1] \to P$ that does not intersect φ. Then we have equality of winding numbers: $W(\varphi, y_0) = W(\varphi, y_1)$.*

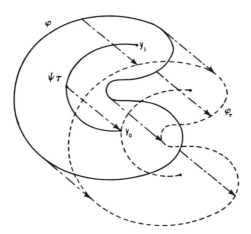

Figure 25.1

For each $t \in [a, b]$ and $\tau \in [0, 1]$, let $\Phi(t, \tau)$ be the point at the end of the vector which starts at φt and is parallel to and of the same length as the vector from $\psi \tau$ to y_0 (see Fig. 25.1). Thus if we set $\varphi_\tau t = \Phi(t, \tau)$ for a fixed τ, the curve φ_τ is obtained by a parallel translation of the curve φ. Think of the plane P as a rigid sheet of metal with a slot cut along the curve ψ. Nail the sheet to the wall by a nail at y_0 of diameter less than the width of the slot. Now slide the sheet along the wall so that the nail follows the course of the slot, but do not allow the sheet to rotate. The resulting motion of the closed curve gives a picture of the homotopy Φ we have constructed.

From the constancy of the winding number under a homotopy, it follows that $W(\varphi_0, y_0) = W(\varphi_1, y_0)$. Moreover the pair (φ_1, y_0) is congruent to the pair (φ_0, y_1) under the translation of the plane by the vector from y_0 to y_1. Since the winding number is obviously unchanged by a congruence, $W(\varphi_1, y_0) = W(\varphi_0, y_1)$. Combining these equalities gives the conclusion of the theorem.

Exercises

1. In Fig. 25.2, show that the winding number of the circle φ_0 about y is the same as the winding number of the congruent circle φ_1 about y by sketching a homotopy to which we may apply Theorem 25.1.

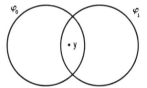

Figure 25.2

2. In Fig. 25.3, show by sketching a homotopy why $W(\varphi_0, y) = W(\varphi_1, y)$; explain why this is not applicable to $W(\varphi_0, x)$ and $W(\varphi_1, x)$.

3. Explain why the homotopy of Fig. 25.4 is improper as a homotopy for Exercise 2 that does not meet x.

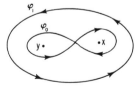 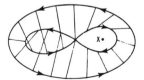

Figure 25.3   Figure 25.4

26. Proof of the main theorem

We have at hand now all the machinery needed to prove our main theorem as stated in Section 18. Suppose y is not in the image fD. Let $\varphi_0: [0, 1] \to C$ be the standard representation of C as a closed curve. Let Φ be the homotopy described in Section 24 which shrinks φ_0 over D to the center point z of D. Then $f\Phi$ is a homotopy of $f\varphi_0$ into the constant closed curve at fz. Since $f\Phi Q \subset fD$, y is not in the image of the homotopy. Therefore $W(f\varphi_0, y) = W(f\varphi_1, y)$ by Theorem 25.1. Since $f\varphi_1$ is a constant curve, $W(f\varphi_1, y) = 0$ (see Section 23). Hence $W(f\varphi_0, y) = 0$. Thus we have proved: If y is not in fD, then $W(f\varphi_0, y) = 0$. Therefore $W(f\varphi_0, y) \neq 0$ implies y is in fD. This completes the proof. (Fig. 26.1 illustrates successive stages of the homotopy $f\Phi$ for the mapping f described in Section 15.)

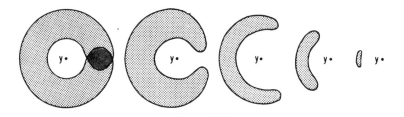

Figure 26.1

Exercise

1. If the disk D in the statement of the theorem is replaced by a rectangle D' and its interior, how must the proof be modified? What replaces the standard representation of φ_0 and the homotopy Φ?

27. The circle winds once about each interior point

In this section, we shall show that, if a mapping of a disk D into a plane leaves fixed each point of its boundary, then all points of D lie in the image of D. To pave the way for the proof of this theorem, we shall first prove something that we accepted in Section 17 as intuitively clear: the circle winds once about each interior point.

LEMMA. *Let C be a circle in the plane, y a point of the interior of C, and let $\varphi_0: [0, 1] \to C$ be the standard representation of C as a closed curve (see Section 16); then $W(\varphi_0, y) = 1$.*

To prove this, we use the partition $\mathcal{P} = \{0, \frac{1}{2}, 1\}$. Recall that φ is defined by choosing a reference point $\varphi 0$, and then each $t \in [0, 1]$ is mapped into the point of C whose angular measure from $\varphi 0$ in degrees is $360t$. Therefore $\varphi[0, \frac{1}{2}]$ is the semicircle from $\varphi 0$ to $\varphi \frac{1}{2}$ in the counterclockwise direction and $\varphi[\frac{1}{2}, 1]$ is the semicircle from $\varphi \frac{1}{2}$ to $\varphi 1 = \varphi 0$ in the counterclockwise direction (see Fig. 27.1). Taking y to be the center, the prescription of Section 22 gives $W(\varphi, y) = r - s$, where $r = 1$ and $s = 0$. Thus the lemma is true for the center. But each interior point can be joined to the center by a line segment not meeting C. Hence, by Theorem 25.2, it has the same winding number as the center.

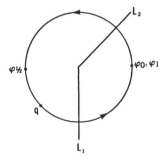

Figure 27.1

As an intuitive formulation of the next proposition, consider the effect of gluing the circular edge of a thin flexible rubber disk to the top of a table. If we want to see what is under the disk, we can achieve nothing by stretching, pulling, or twisting the rubber as long as the edge remains fixed.

THEOREM 27.1. *Let $f: D \to P$ be a mapping of a disk into a plane which leaves fixed each point of its bounding circle C; then the image fD contains all of D.*

Let $\varphi_0 : [0, 1] \to C$ be the standard representation of C as a closed curve, and let y be an interior point of D. Since f leaves fixed each point of C, $f \mid C$ is the identity map, that is $f\varphi_0 = \varphi_0$. Therefore $W(f\varphi_0, y) = W(\varphi_0, y)$. The preceding lemma asserts that $W(\varphi_0, y) \neq 0$; hence $W(f\varphi_0, y) \neq 0$. The main theorem is applicable now and asserts that $y \in fD$. Each point of C is in D since $fC = C$. Thus all points of D are in fD.

As a consequence of this theorem, we have the

COROLLARY 27.2. *There is no continuous mapping of a disk D into its boundary C which leaves fixed each point of C.*

One can, of course, map a rectangle and its interior onto one of its bounding edges so that this edge remains fixed. Just picture rolling up a window shade; under this mapping f, $fx = x$ for all points x on the roller. The edge represented by the roller is said to be a *retract* of the rectangular region (the shade). The corollary says that we cannot roll up the interior of a circle onto its boundary; that is, a circle is not a retract of the disk. Picture the circle and its interior as the rim and membrane of a drumhead. The assertion is that the entire membrane cannot be stretched and rolled onto the rim. Because of this visualization, the corollary is sometimes referred to as the drumhead principle.

Exercises

1. Show that the periphery E of a rectangular region F is not a retract of F; state the theorem, corollary and proofs corresponding to these. Hint: Use a homeomorphism $h: P \to P$ which maps a disk D onto the region F (constructed as in Section 14, the center z of D is carried into the center hz of F and each ray issuing from z is mapped by a similarity onto the parallel ray issuing from hz).

2. Let y be a point on the boundary C of a disk D; show that there is a continuous mapping of $D - y$ onto $C - y$ which leaves fixed each point of $C - y$.

3. If y_0 is a point of the interior of D, show that $D - y_0$ can be retracted continuously onto C.

4. Show that each of the following can be a retract of the disk D:

    (a) any diameter of D;   (b) any one point of D.

5. State and prove the analog in dimension 1 of Corollary 27.2.

28. The fixed point property

In Section 9, Part I, we proved that a continuous mapping of a line segment into itself has at least one fixed point. We prove now the analogous theorem for a disk.

THEOREM 28.1. *Let $f: D \to D$ be a mapping of a disk into itself; then f leaves fixed at least one point of D, that is, $fx = x$ for at least one point $x \in D$.*

FIXED POINT PROPERTY

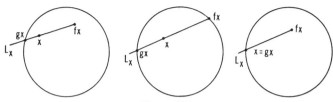

Figure 28.1

Let us assume, to the contrary, that there is a map $f: D \to D$ without any fixed point. Then for each $x \in D$, fx and x are distinct. Hence we may construct a ray L_x issuing from fx and passing through x. Let gx be the point of C in which the ray L_x meets C. In case fx happens to be on C, gx is the other point in which L_x meets C; and if $x \in C$, then $gx = x$. Fig. 28.1 shows several of the possibilities. Thus $g: D \to C$, and $g \mid C$ is the identity. We shall prove that g is continuous; then g will contradict Corollary 27.2, and this contradiction will prove our theorem.

To prove the continuity of g, let $x_0 \in D$ and let V be a neighborhood of gx_0. We shall construct a neighborhood U of x_0 such that $x \in U$ implies $gx \in V$. Let b, c denote the endpoints of V, and let m be the midpoint of the segment x_0 to fx_0. Let H be the line through b and m, and let K be the line through c and m. Choose a circular neighborhood N of fx_0 containing no points of H or K. Since f is continuous, there is a circular neighborhood U' of x_0 such that $fU' \subset N$. Choose now a neighborhood U of x_0 such that U contains no points of H or K and $U \subset U'$. We have then the situation pictured in Fig. 28.2: U is a neighborhood of x_0, N is a neighborhood of fx_0, neither U nor N meet H or K, and $fU \subset N$. Moreover, a point $x \in U$ and its image $fx \in N$ lie on opposite sides of H (and also of K) because x_0 and fx_0 lie on opposite sides of H (and also of K) and U and N are connected sets not meeting H (and K).

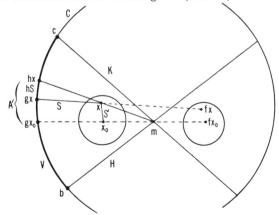

Figure 28.2

Hence L_x, the ray from fx through x, meets both H and K between fx and x. Therefore the segment S of L_x leading from x to gx contains no points of H or K. Let h denote the radial projection from m onto C. Since m is on L_{x_0}, h sends x_0 into gx_0. Now h maps the segment S' from x_0 to x into an arc A' of C starting at gx_0. Since S' does not meet H or K, A' cannot contain b or c. Hence A' lies wholly in V, and so $hx \in V$. Since S does not meet H or K, it follows as before that $hS \subset V$. The radial projection of gx from m onto C is gx; that is, $hgx = gx$. Taken together, the three statements $gx \in S$, $gx = hgx$, and $hS \subset V$ imply that $gx \in V$, and this completes the proof of the continuity of g.

Exercises

1. Let D be a disk with center z and radius r. Find the fixed point in each of the following mappings of D into D:

   (a) a rotation about the center,

   (b) a reflection in a diameter,

   (c) a contraction similarity toward the center,

   (d) a contraction to half size followed by a translation of $\frac{1}{3}r$,

   (e) a reflection in a vertical diameter, followed by a contraction toward z to half its size, followed by a translation of $\frac{1}{3}r$ to the right.

   (f) Each ray from z is mapped onto a ray from z forming an angle with the horizontal which is twice the original angle with similarity ratio $\frac{1}{2}$, and translated $\frac{1}{3}r$ to the left.

2. Show that, if E is homeomorphic to D, then any mapping $E \to E$ has a fixed point.

29. Vector fields

A vector in a plane or in space is an ordered pair of points. It is customary to picture the vector as the line segment connecting the two points, oriented from the first point to the second by an arrowhead. The algebraic properties of vectors make them indispensable tools for the study of the geometry of euclidean spaces of higher dimension. Vectors are of utmost importance in mathematical physics where they are used to represent forces, velocities, and accelerations.

We shall use the concept of velocity vector to obtain an intuitive appreciation of the theorems to be proved in the following sections. If a moving point passes through a point x, its *velocity vector* at x is the vector v from x whose direction is the instantaneous direction of motion, and whose length is the instantaneous speed. If the point were to continue to travel in the same direction at the same speed for one unit of time, it would arrive at the endpoint x' of v. Constant velocity (i.e. constant direction and constant speed) is just the simplest case, but we must consider motions of points along a curve. In such a case the direction and speed of the point will ordinarily change as it moves along the curve. The velocity vector at each point of the curve is tangent to the curve, oriented in the same direction as the curve, and its length is the instantaneous speed (rate of eating up arclength). For example, the points labeled 0, 1, 2, 3, ··· in Fig. 29.1 indicate the various positions at equal time intervals of a particle traversing a curve, and tangential vectors have been attached to each point to indicate the velocity of the particle at these points. Thus the smaller velocity vectors at 1 and 2 accord with the short distance from 1 to 2 and from 2 to 3; a spurt in speed is indicated at 3, a slowing down at 4, and high speed at 5, etc.

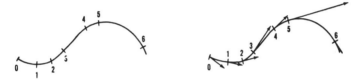

Figure 29.1

A *vector field* v is a function which to each point x of a region of a plane (or space) assigns a vector vx issuing from x. If one considers a flowing liquid or gas, the velocity vectors of the various particles at a single instant of time form a vector field. For example, under a steady flow in a fixed direction at a fixed velocity, the vectors are all parallel and of equal lengths. As another example, consider a rotation about a point z in the plane at constant angular velocity (see Fig. 29.2). At a point x, vx is perpendicular to the line through z and x, and its length is proportional to $d(z, x)$. The vector vz is the ordered pair (z, z); it has no direction, and its length is zero. Such a vector is called the *zero vector*.

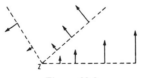

Figure 29.2

A flowing fluid whose velocity field is the same at all times is called a *steady* flow. Precisely, the velocity vector depends only on the location of the particle in the plane (or space) and does not depend on the time.

Two particles passing through the same point at different times have the same velocity vectors at that point. The two examples discussed above are examples of steady flows. Such flows have *streamlines*. These are the paths of the particles. They form a family of curves such that one and only one curve passes through each point, and the velocity vector at the point is tangent to the curve. The particles along any streamline remain on the streamline. One can think of the streamline as sliding along itself. In the first example above, the streamlines form a family of parallel lines. In the second example, they form the family of circles with center z. The streamline through z is the constant curve at z.

30. The equivalence of vector fields and mappings

At first glance a vector field may seem a rather difficult concept to handle mathematically. However it is entirely equivalent to the concept of a mapping in the following way. Suppose $f: A \to P$ is a mapping of a subset A of the plane into the plane. Let o be a fixed reference point of P called the *origin*. For each $x \in A$, let vx be the vector issuing from x which is parallel to the vector from o to fx and has the same length. Thus to each mapping f is assigned a vector field v. Conversely, if a field v is given on A, we can define f by saying that fx is the endpoint of the vector issuing from o which is parallel and equal in length to vx. It is readily seen that this correspondence between vector fields and mappings is one-to-one.

This correspondence can be clarified by using the concept of the equivalence of two vectors. Two vectors are called *equivalent* vectors if they are parallel, have the same length, and are similarly oriented. In case one of them is the zero vector, they are equivalent only when both are zero. Now if v is a vector field, the vector vx based at x is equivalent to a unique vector based at o, and this vector is uniquely determined by its endpoint fx. Conversely, if $f: A \to P$ is a mapping, we obtain the corresponding vector field v by defining vx to be the vector based at x which is equivalent to the vector from o to fx.

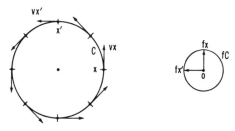

Figure 30.1

The left diagram of Fig. 30.1 pictures the vector field on a circle C of radius r consisting of the tangent vector at each point, of length $\frac{1}{2}r$, oriented counter-clockwise. The right diagram of Fig. 30.1 pictures the image of the associated mapping. In this case f shrinks C by a similarity to a circle half its size with center o, and rotates it through 90°. Fig. 30.2 illustrates a field of outward normals of lengths $\frac{1}{2}r$. In this case f shrinks the circle to half its size, but does not rotate it. The field of inward normals of the same length would give an f obtained from the preceding by a 180° rotation about o.

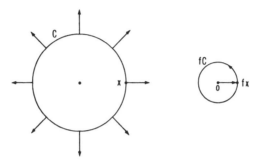

Figure 30.2

In case f is a constant function sending all of A into a single point, the image fx is the same point for all $x \in A$. Then the ordered pair (o, fx) is unique for all $x \in A$ and all of the vectors of the corresponding field are parallel and of the same length and orientation. Such a field is called a *constant field*.

The one-to-one correspondence between vector fields and mappings is used to carry over to vector fields concepts and properties defined for mappings. For example, a vector field v is called *continuous* if the corresponding function f is continuous.

Exercise

1. For each of the following mappings of P into P, write a description or draw a picture of the corresponding vector field:

   (a) f maps all the points of P into a single point not the origin;

   (b) f is the identity mapping;

   (c) f is a 180° rotation about the origin;

   (d) f translates all points in a fixed direction by a fixed distance;

   (e) f is the reflection in a line through o.

31. The index of a vector field around a closed curve

Let v denote a continuous vector field on a subset A of the plane P, and let $\varphi \colon [a, b] \to A$ be a closed curve in A. Think of the curve as a moving point. At each position x on the curve, the vector vx is defined. As the point traverses the curve, the vector will vary continuously, rotating its direction and changing its length. When the point returns to the initial position, the vector must return to its initial direction and length. One can ask how many complete rotations of the direction were made by the vector as the point traversed the curve. The question and its answer are most clearly formulated by means of the associated mapping $f \colon A \to P$ described in Section 30. The composition $f\varphi \colon [a, b] \to P$ is a closed curve in the plane. If the origin o is not on $f\varphi$, then the winding number $W(f\varphi, o)$ is defined. We shall also call it the *index* of the vector field v around the closed curve φ and denote it by $I(v, \varphi)$. Thus

$$I(v, \varphi) = W(f\varphi, o).$$

In the example of Fig. 30.1 we clearly have $I(v, C) = 1$. The same is true of the example of Fig. 30.2, and also of the third example of the inward normals. However a constant field, as in Fig. 31.1, has index zero. A field of index 2 on a circle is shown in Fig. 31.2.

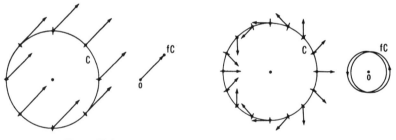

Figure 31.1   Figure 31.2

THEOREM 31.1. *Let v be a continuous vector field defined on a disk D in the plane and such that vx is not the zero vector for any point x on the boundary circle C of D. If the index of v around C, $I(v, C)$, is not zero, then there is at least one point x in D whose vector vx is zero.*

This theorem is just a translation from "mapping" language into "vector field" language of our main theorem of Part II. Let $f \colon D \to P$ be the mapping which corresponds to the field v, and let $\varphi \colon [0, 1] \to C$ be the standard representation of C as a closed curve. Then $W(f\varphi, o) = I(v, \varphi)$ is not zero by hypothesis. Our main theorem asserts that the equation $fx = o$ has at least one solution x. Then the corresponding vx must be the zero vector for it is equivalent to the vector from o to o.

THEOREM 31.2. *Let v be a continuous field of non-zero vectors defined on a disk D. Then on the boundary C of D, there is at least one point x where vx is an outward normal, at least one point x' where vx' is an inward normal, and there are at least two points on C where the vectors are tangent to C. In general, for each angle α, there is at least one point $x \in C$ such that vx and the outward normal at x form the angle α.*

The conclusions rest heavily on the assumption that v has no zeros inside D. For example, let D have center o, let $f: D \to P$ be the identity mapping. The corresponding v has just one zero at the center of D. The conclusion of the theorem is not true here; for every $x \in C$, vx is the outward normal at each point. A good illustration of the theorem is provided by the constant field (Fig. 31.1); each α is taken on exactly once.

It suffices to prove the last conclusion of Theorem 31.2, for it implies the preceding ones; just let α be the angle of $0°$ for the outward normal, $180°$ for the inward normal, and $90°$, $270°$, for the tangents.

To prove the last conclusion, let α be fixed. Choose the origin o at the center of D, and let $f: D \to P$ be the mapping which corresponds to the field v. Let g denote the rotation of P about o through the angle $-\alpha$. Let h denote the radial projection from o onto C. Since v is never zero, fD and gfD do not contain o. Hence $hgf: D \to C$ is defined. Since $C \subset D$, hgf can be regarded as a mapping $D \to D$. Theorem 28.1 asserts that hgf leaves fixed at least one point; that is, there is an $x \in C$ such that $hgfx = x$. Let $v'x$ be the vector which corresponds to x under the mapping hgf. By definition it is parallel to the vector from o to $hgfx = x$. Therefore $v'x$ is the outward normal at x. But for any point y of D, how do vy and $v'y$ differ? Since hgf is obtained by applying first g and then h to f, it follows that $v'y$ is obtained from vy by rotating it about y through an angle $-\alpha$ and then changing its length to the radius of C. Thus, at the fixed point x on C, the vector vx must make the angle α with $v'x$ which has been shown to be the outward normal.

COROLLARY. *If v is a continuous vector field on a disk D, and if, on C, v is never tangent (never normal) to C, then v has at least one zero in D.*

The foregoing results are of significance in the study of steady flows. A zero of the velocity field occurs only at a point which remains fixed during the flow. Suppose a velocity field on a disk is such that, on its boundary C, the field is the inward normal. Clearly the fluid flows into D at each point of C. Intuition says that the fluid must pile up somewhere inside D. Since the field is nowhere tangent to C, the preceding corollary gives us at least one fixed point of the flow where fluid can congest.

Exercises

1. Let D be a disk with center z. For each point x of D, let vx be the vector of a fixed length s, oriented along the ray from z to x. Then on the boundary C, each vx is an outward normal, there is no inward normal, and there is no point at which vx is tangent to C. Does this contradict the conclusions to the second theorem of this section?

2. For each of the following mappings of the disk D, find the index of the field on C, find the points $x \in D$ where vx is zero, the points $x \in C$ where vx is tangent to C, where it is an outward normal, and where it is an inward normal to C.

    (a) f maps all points of D into the center z (and $o \neq z$);

    (b) f is the identity mapping and o is at the center z;

    (c) f is the identity mapping and o is at distance $r/2$ from z;

    (d) f is the identity mapping and $d(o, z) > r$;

    (e) f is a 180° rotation about the center, and o is an exterior point of D;

    (f) f translates by a fixed vector and o is exterior to fD;

    (g) f is the reflection about a chosen diameter and o is at z.

32. The mappings of a sphere into a plane

By a sphere S we shall mean the set of all points in space whose distance from a point z (the center) is a fixed positive number r (the radius). If $x \in S$, the *antipode* x' of x is the other point in which the line through x and z meets S, i.e., the diametrically opposite point.

If we map the sphere S into a plane P by a perpendicular projection f, then there is a pair of antipodes x, x' for which $fx = fx'$, namely, the intersections of S with the line through z perpendicular to P. It is a surprising fact that a part of this result holds true for any mapping of S into P. The following analog of Theorem 10.1 was discovered by the mathematicians K. Borsuk and S. Ulam in 1933.

THEOREM 32.1. *Every mapping $f: S \to P$ of a sphere into a plane maps some pair of antipodes of S into the same point; that is, for at least one pair x, x' of antipodes, $fx = fx'$.*

To prove the theorem, choose a point o in P as origin. For each $x \in S$ define gx in P to be the endpoint of a vector issuing from o equivalent to the vector from fx to fx', where x' is the antipode of x (see Fig. 32.1). Thus g is also a mapping $g: S \to P$. It has the property that, for every $x \in S$, gx' is symmetric to gx about o, because x is the antipode of x', and, to construct gx', we have only to reverse the arrow from fx to fx'. The theorem may now be proved by showing that g maps some point of S onto o, since it is only then that the vector and its opposite coincide.

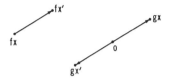

Figure 32.1

Assume as known that g is continuous; this will be proved later. Let P' be a fixed plane through the center z of S, let C denote the circle in which P' meets S, and let D be the disk in P' whose boundary is C. In case the image gC contains o, there is an $x \in S$ such that $gx = o$, and the theorem is proved. So we need consider only the case where gC does not contain o. Let H be one of the hemispheres into which S is divided by P', and let $\psi: D \to H$ be the inverse of the perpendicular projection of H onto D. Then the composition $g\psi: D \to P$ coincides with g on the boundary C. Let φ be the standard representation of C as a closed curve (Section 16). It will suffice now to prove that the winding number $W(g\varphi, o)$ is not zero, because, once this is done, the main theorem (Section 18) assures us of the existence of a point y in D such that $g\psi y = o$, and this y yields the point $x = \psi y$ in S that satisfies $gx = o$.

We shall prove that $W(g\varphi, o) \neq 0$ by showing that $W(g\varphi, o)$ is an odd integer (recall that zero is even because $2 \cdot 0 = 0$). Let φ_1 and φ_2 be the restrictions of φ to the subintervals $[0, \frac{1}{2}]$ and $[\frac{1}{2}, 1]$. Let $x_0 = \varphi 0 = \varphi 1$; then its antipode $x'_0 = \varphi \frac{1}{2}$. Moreover φ_1 represents one semicircle of C as a curve from x_0 to x'_0, and φ_2 represents the other from x'_0 to x_0. Select now a partition of $[0, \frac{1}{2}]$ which is sufficiently fine for the curve $g\varphi_1$ relative to o, and apply the result of Section 22 for computing $A(g\varphi_1, o)$. It states that

$$A(g\varphi_1, o) = u - v + (r - s)360,$$

where $r - s$ is an integer, and u and v are the protractor readings for the rays from o to gx'_0 and gx_0 respectively. Since x_0, x'_0 are antipodal, the points gx'_0, o, gx_0 are in a straight line; hence $u - v$ is the measure in degrees of a straight angle, that is, $u - v = \pm 180$. It follows that $A(g\varphi_1, o)$ is an *odd* multiple of 180:

$$A(g\varphi_1, o) = (2m + 1)180.$$

Fig. 32.2 illustrates the case where the multiple is -3.

Consider next the curve $g\varphi_2$ from gx'_0 back to gx_0. Let $h: P \to P$ be the rotation about o through $180°$. If $t \in [0, \frac{1}{2}]$, then $t + \frac{1}{2} \in [\frac{1}{2}, 1]$, and $\varphi_2(t + \frac{1}{2})$ is the antipode of $\varphi_1 t$. It follows now from the symmetry property of g that

$$g\varphi_2(t + \tfrac{1}{2}) = hg\varphi_1 t \quad \text{for all } t \in [0, \tfrac{1}{2}].$$

That is to say, the curve $g\varphi_2$ is obtained by rotating the curve $g\varphi_1$ through $180°$ (see Fig. 32.2). Since a rotation preserves angles, $A(g\varphi_1, o) = A(g\varphi_2, o)$. Using the additive property of A, we obtain

$$A(g\varphi, o) = A(g\varphi_1, o) + A(g\varphi_2, o) = 2A(g\varphi_1, o)$$
$$= 2(2m + 1)180 = (2m + 1)360,$$

hence

$$W(g\varphi, o) = A(g\varphi, o)/360 = 2m + 1.$$

This completes the proof that $W(g\varphi, o)$ is odd.

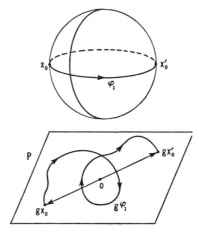

Figure 32.2

It remains to show that g is continuous. Let x_0 be a point of S, and let N be a circular neighborhood of gx_0 of radius r. Let U, U' be circular neighborhoods of fx_0, fx'_0, respectively, each of radius $r/2$. Since f is continuous, there are neighborhoods V, V' of x_0, x'_0, respectively, such that $fV \subset U$ and $fV' \subset U'$. The set T of antipodes of points of V' is a neighborhood of x_0; let W be a neighborhood of x_0 contained in both V and T. Then if x is a point in W, we have $x \in V$ and $x' \in V'$; so it follows that $fx \in U$ and $fx' \in U'$. Let y be the point in P such that the vector from fx_0 to y is equivalent to the vector from fx to fx' (see Fig. 32.3). Since by definition the vector from o to gx is also equivalent to the vector from fx to fx', and the vector from o to gx_0 is equivalent to the vector from fx_0 to fx'_0, one sees that the distance

$$d(gx, gx_0) = d(y, fx'_0).$$

By the triangle inequality, this is in turn less than or equal to

$$d(y, fx') + d(fx', fx'_0).$$

From the parallelogram, $d(y, fx') = d(fx_0, fx)$, and both this distance and $d(fx', fx'_0)$ are less than $r/2$; so their sum is less than r. This implies $gx \in N$. Hence g maps W into N. This proves the continuity of g, and thus completes the proof of the theorem.

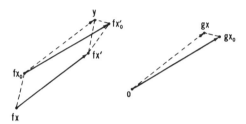

Figure 32.3

As an application, let us assume that the surface of the earth is a sphere S, and that at any instant of time, the air pressure px and temperature tx are continuous functions of $x \in S$. In a plane P, choose a cartesian coordinate system by selecting an origin, two perpendicular oriented lines through it, and a unit of measure. For each $x \in S$ let fx be the point of P whose coordinates are (px, tx). Since p and t are continuous, it follows that $f: S \to P$ is continuous. We now apply our theorem to this f and obtain the

COROLLARY. *At each instant of time, there is a pair of antipodal points on the earth's surface where the pressures and also the temperatures are equal.*

Clearly the physical properties of pressure and temperature have nothing to do with the conclusion; p and t can be any two continuous real-valued functions defined on S.

Notice also that if we consider just a single function, say, temperature, then Theorem 10.1 tells us that on each great circle there is a pair of antipodes where the temperatures are equal.

Exercises

1. Let S denote a sphere in R^3 of radius r with center z at the origin of the coordinate system (x_1, x_2, x_3). Let P denote the (x_1, x_2)-plane, and let L denote the x_1-axis. Find the pairs of antipodes with the same image

under $f: S \to P$ if f is (a) the perpendicular projection of S into L (all projecting rays are perpendicular to L), (b) the composition of the 90° rotation of S about L followed by the perpendicular projection into P.

2. Show that the conclusion of the Borsuk–Ulam theorem is still true if "antipodes" are defined using lines through a point Q inside S other than the center.

3. Show that the conclusion of the theorem is true if the sphere is replaced by an ellipsoidal surface, or the surface of a rectangular box, and antipodes are symmetric with respect to the center.

33. Dividing a ham sandwich

The theorem of this section is the three-dimensional analog of Theorem 11.1 which says that any pair of bounded, connected regions in the plane can be divided exactly in half (in the sense of area) by a single line.

THEOREM 33.1. *Let A, B, C be three bounded and connected open sets in space. Then there is a single plane which divides each exactly in half by volume.*

An illustration is provided by three spherical balls and the plane through their centers. The strength of the theorem is that it applies even when the regions are irregular. If we interpret A, B to be slices of bread, and C to be a slice of ham between them, then the conclusion can be interpreted: With one stroke of a knife, a ham sandwich can be divided so that both slices of bread and the ham are cut exactly in half.

To prove the theorem, we choose a sphere S which encloses A, B and C. There is such a sphere because A, B and C are bounded. Let z be the center of S and r its radius. For each $x \in S$, let L_x denote the diametrical line through z and x. We shall show that:

(1) For every $x \in S$, there is a unique point x_A on L_x such that the plane perpendicular to L_x at x_A divides A in half by volume.

Once this is done, we let $g_A x$ be the distance $d(z, x_A)$ with a positive sign if x_A is on the segment from z to x, and with a negative sign if x_A is on the segment from z to the antipode x' of x. Since L_x and $L_{x'}$ coincide and have oppositely oriented coordinates, and since $x'_A = x_A$, it follows that $g_A x' = -g_A x$.

We define x_B, $g_B x$ and x_C, $g_C x$ in the analogous fashion, using B and C in place of A. Now consider the mapping $f: S \to R^2$ which assigns to each $x \in S$ the point with the coordinates

$$fx = (g_A x - g_B x,\ g_A x - g_C x).$$

We shall show that:

(2) f is continuous.

Once we have proved (1) and (2), the proof of Theorem 33.1 is concluded very quickly as follows. By Theorem 32.1, there is a point x such that $fx = fx'$. Equating the coordinates of fx and fx' yields

$$g_A x - g_B x = g_A x' - g_B x'$$
$$g_A x - g_C x = g_A x' - g_C x'$$

With the help of the relations $g_A x' = -g_A x$, $g_B x' = -g_B x$, and $g_C x' = -g_C x$ noted above, the first equation reduces to $g_A x = g_B x$, and the second to $g_A x = g_C x$. Hence, for the point $x \in S$ whose image coincides with that of its antipode, we have $x_A = x_B = x_C$, and the plane perpendicular to L_x through this point divides all three regions in half.

To prove (1), for each point $y \in L_x$, let P_y denote the plane through y perpendicular to L_x, and let hy denote the volume of the part of A on the same side of P_y as x. As y varies from x' to x, hy varies from the volume of A down to zero. If y_1, y_2 are in L_x, the difference $|hy_1 - hy_2|$ is at most the volume of the part of the solid sphere between the planes P_{y_1} and P_{y_2}, and this is less than $\pi r^2 |y_1 - y_2|$. This shows that h is continuous at each point y of L_x (with $\delta = \epsilon/\pi r^2$). So the main theorem of Part I assures us that there is a point y such that hy is half the volume of A. If there were two such points, there would be two parallel planes P_1 and P_2 dividing A in half. The slab Q of space between P_1 and P_2 separates the rest of space into two disconnected parts. Since A is connected and has half its volume in each part, A must contain a point q inside Q. Since A and Q are open, there is a spherical neighborhood U of q contained in $A \cap Q$. Since U has a positive volume, so does $A \cap Q$. Shifting from P_1 to P_2 alters hy by the volume of $A \cap Q$; so both P_1 and P_2 could not divide A in half. This proves (1).

To prove (2), the continuity of f, it suffices to prove the continuity of each coordinate of f. We shall prove only that g_A is continuous, and leave the rest to the reader. Let c be a point of S at which the continuity of g_A is to be proved, and let c_A be the point on L_c where the plane P_c perpendicular to L_c cuts A in half. Let x be a point on S near c, and let x_A, P_x be defined similarly. Fig. 33.1 shows the intersection of the configuration with the plane through c, x, z. We want to show that $|g_A x - g_A c|$ can be made small (less than a prescribed $\epsilon > 0$) by restricting x to be near to c (in $N(c, \delta)$).

Let u and v be the points where the great circle through c and x meets P_c. Let P' and P'' be the planes perpendicular to L_x passing through u and v respectively. Let N denote the interior of S. The part of N on the same side of P' as c is included in the part of N on the same side of P_c as c. Since $A \subset N$, the part of A on the same side of P' as c is included in the part of A on the same side of P_c as c. So if V denotes the volume of A, the volume of the part of A on the same side of P' as c is at most $V/2$. By a similar argument, the volume of the part of A on the side of P'' opposite to c is at most $V/2$. It follows that P_x must lie between P' and P''. Therefore, if w is the distance between P' and P'', we have

$$|g_A x - g_A c| < w.$$

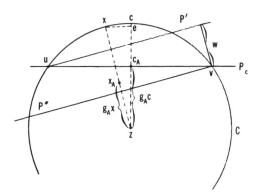

Figure 33.1

An estimate of the size of w is obtained from noting from the similarity of two triangles that

$$\frac{w}{d(u, v)} = \frac{d(e, x)}{d(z, x)}.$$

where e is the perpendicular projection of x on L_c. Since $d(z, x) = r$, this gives

$$w = \frac{d(u, v)}{r} d(e, x).$$

Since $d(u, v) \leq 2r$, and $d(e, x) \leq d(c, x)$, we obtain

$$w \leq 2 d(c, x).$$

Therefore

$$|g_A x - g_A c| < 2 d(c, x).$$

For a given $\epsilon > 0$, take $\delta = \epsilon/2$; then $x \in N(c, \delta)$ implies that $|g_A x - g_A c| < \epsilon$. This shows that g_A is continuous at c. Since this holds for each $c \in S$, it follows that g_A is continuous on S. The proof of Theorem 33.1 is now complete.

Exercises

1. If A is a solid spherical ball, B is a solid cube, and C is a solid cylinder, describe the plane which cuts all three in half.

2. Give a direct proof of the theorem in case A is a solid spherical ball, B is a solid hemisphere whose axis passes through the center of A, and C is any third solid.

34. Vector fields tangent to a sphere

Let v denote a vector field defined on a sphere S in space (see Section 29). It assigns to each point x of S an oriented line segment issuing from x. We shall say that the field v is *tangent* to S if, for each x of S, the line segment issuing from x is tangent to S, or equivalently, if it is perpendicular to the radial line zx where z is the center of S. As in Section 30, we associate with v a mapping g of S into space by choosing an origin o and defining gx to be the endpoint of the vector issuing from o parallel and equal in length to vx. We say that v is *continuous* whenever the associated g is continuous.

THEOREM 34.1. *Let v be a continuous vector field defined over a sphere S and tangent to S. Then there is at least one point x of S such that $vx = 0$.*

A vector field tangent to S can be interpreted as a flow. The theorem then implies that any steady flow on a spherical surface has at least one stationary point. To give this a practical aspect, assume that the earth's surface is a sphere, and that the velocity vector of wind flow is continuous. Then, at any instant of time, there is some place on earth where the wind is not blowing.

If we rotate a sphere about an axis at a constant angular velocity, we obtain a flow having two stationary points.

To illustrate the theorem, let us construct a tangent field to S having exactly one zero at a point x_0. Let L be an oriented tangent line to S through x_0. For any $x \in S$ distinct from x_0, x and L determine a plane P_x which intersects S in a circle C_x. Give to C_x the same orientation as that of its tangent line L. Define vx to be the vector in P_x issuing from x tangent to C_x, whose length is half the distance $d(x, x_0)$, and which is oriented concordantly with C_x. Fig. 34.1 shows several of the vectors tangent to C_x in P_x; we have oriented L upwards, and then C_x counterclockwise. Notice that the vectors become shorter and shorter as x approaches x_0. We complete the definition of v by setting $vx_0 = 0$ and obtain a continuous field.

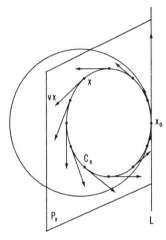

Figure 34.1

In contrast to a sphere, the surface of a torus (an inner tube) does possess continuous tangent vector fields which are nowhere zero. Picture the velocity field of an inner tube rotating about an axle. Equally well, picture a smoke ring.

PROOF. Choose a fixed plane P through the center z of S. Let C denote the circle $P \cap S$, and let D be the disk it bounds in P. Denote by H and H' the two closed hemispheres of S determined by C. Let p and p' be the poles where the line through z perpendicular to P meets H and H' respectively. Let $h: H' \to D$ be the topological equivalence given by stereographic projection from p. Precisely, if $x \in H'$, then hx is the intersection of D with the segment p to x. Similarly, let $h': H \to D$ be given by stereographic projection from p'.

It suffices to prove: if v has no zero vector on one of the hemispheres, say H', then it must have a zero vector on H. As the first major step, we shall prove:

1. The stereographic projection h maps the field v on H' into a field w' on D so that vectors at corresponding points have the same lengths. Similarly, h' projects the field v on H into a field w on D.

Once this is done, the fact that v has no zero vectors on H' will imply that w' has no zero vectors on D, hence the index $I(w', C) = 0$ by Theorem 31.1. Then, as the second major step, we shall prove

2. At a point x of C the vectors wx and $w'x$ are obtained by rotating vx about the tangent line to C, first 90° one way and then 90° the other. This fact and $I(w', C) = 0$ imply that $I(w, C) = 2$.

Once this is proved, we apply Theorem 31.1 and $I(w, C) \neq 0$ to conclude that w has a zero vector at some point of D; hence v has a zero vector at the corresponding point of H. It remains therefore to prove 1 and 2.

For each point x of S, let T_x denote the plane through x tangent to S. When x is in H', define a mapping $h_x: T_x \to P$ by parallel projection using the family of lines parallel to the line through p and x. Precisely, if $q \in T_x$, then $h_x(q)$ is the point of intersection of P with the line through q parallel to the line through p and x (see Fig. 34.2). It is an exercise of elementary geometry to show that T_x and P make equal angles with the line from p to x. Therefore h_x is an isometry (preserves distances). We define the vector field w' on D by defining the vector $w'y$, based at $y \in D$, to be the image in P of the vector vx in T_x under h_x, where x is the point of H' such that $hx = y$. Since h_x is an isometry, the vectors $w'y$ and vx have equal lengths. Similarly, when $x \in H$, we define $h'_x: T_x \to P$ using parallels to the line through p' and x, and wy is the image of vx under h'_x when $y = h'x$. This defines the fields w and w' on D, and completes the proof of 1.

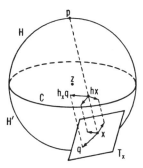

Figure 34.2

When $x \in C$ we have $hx = x$ and $h'x = x$. Moreover the planes T_x and P are inclined at an angle of $45°$ to the line through p and x. Therefore the mapping h_x of T_x into P may be regarded as the result of a $90°$ rotation of T_x about the line L_x tangent to C at x. Similarly h'_x is the result of a $90°$ rotation of T_x about L_x but in the opposite direction. Therefore the vector wx is obtained from $w'x$ by reflecting the plane P in the tangent line L_x to C at x.

As shown in Section 30, the field w' on D may be interpreted as a mapping $f: D \to P$. Since w' has no zero vectors, fD does not contain the origin. The standard shrinking of C over D into the center point of D when composed with f gives a homotopy of the closed curve $f \mid C$ to a constant (see Section 25). Each stage of the homotopy, corresponding to a $\tau \in [0, 1]$, reinterprets as a vector field w'_τ on C, all vectors lying in P. As τ varies from 0 to 1, we obtain a moving vector field on C; that is, for a fixed $x \in C$, the vector $w'_\tau x$ rotates about x as τ varies from 0 to 1. When $\tau = 0$, the field w'_0 is w', and when $\tau = 1$, the field w'_1 is constant (all vectors are parallel and of the same length). Since fD does not contain the origin, none of the vectors $w'_\tau x$ are zero.

Define $w_\tau x$ to be the vector obtained from $w'_\tau x$ by reflection of P in L_x. Then, for each τ, w_τ is a vector field on C with vectors in P, and as τ varies from 0 to 1 we obtain a homotopy of the field $w = w_0$ into a field w_1. Since $w_\tau x$ is never zero, it follows that w and w_1 have the same index on C. The index $I(w_1, C)$ is readily computed by inspecting Fig. 34.3. The constant field w'_1 on C is shown with solid vectors. The field w_1 is shown with dotted vectors. At each point x of C the dotted vector $w_1 x$ is obtained by reflecting the solid vector $w'_1 x$ in the tangent line L_x. Suppose we start at the top of C, in Fig. 34.3, and run once around C in the clockwise direction. At the start $w_1 x = w'_1 x$ points to the right. By the time we have traversed a fourth of C, $w_1 x$ has rotated through 180° counterclockwise, and points to the left. As we continue around C, it continues to rotate at the same speed. Therefore it rotates through 720° counterclockwise as we move once around C, so $I(w_1, C) = 2$. Since $I(w, C) = I(w_1, C)$ it follows that $I(w, C) = 2$. This completes the proof of the theorem.

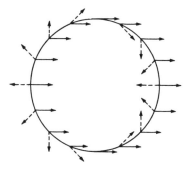

Figure 34.3

Exercises

1. Show that a continuous field of non-zero vectors defined on the sphere (but not required to be tangent) must have at least one vector perpendicular to the sphere.

2. Show that the theorem holds if the sphere is replaced by an ellipsoid (or any smooth ovoid).

35. Complex numbers

It is a familiar fact that some polynomials in a real variable x, for example, $x^4 + x^2 + 1$, have no real zeros. The simplest such polynomial, $x^2 + 1$, led to the introduction of the pure imaginary number $\sqrt{-1}$ which we denote by i. It was then discovered that the zeros of other polynomials could be expressed in the form $a + ib$ where a and b are real numbers. For example, $x^4 + x^2 + 1$ has as zeros

$$\frac{1}{2} + i\frac{\sqrt{3}}{2}, \quad \frac{1}{2} - i\frac{\sqrt{3}}{2}, \quad -\frac{1}{2} + i\frac{\sqrt{3}}{2}, \quad -\frac{1}{2} - i\frac{\sqrt{3}}{2},$$

as can be verified by direct substitution.

This enlargement of the number system should be compared with the various enlargements discussed in Section 5. The need for it is entirely analogous: the real number system is not adequate for solving polynomial equations. Two questions arise immediately: How big a class of numbers must one use so that every polynomial has a zero? How are the new numbers to be pictured geometrically? In the next section we shall show that the set of complex numbers is big enough. In this section we shall review the basic properties of the complex numbers and their geometric interpretations.

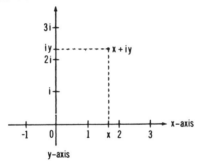

Figure 35.1

Just as the real numbers can be pictured as the points of a line, the complex numbers can be represented by the points of a plane P. A complex number $x + iy$ is just a pair (x, y) of real numbers. Having chosen an origin, two perpendicular coordinate lines, and a unit of length in P, the pair (x, y) can be plotted as the point with the coordinates (x, y), as illustrated in Fig. 35.1. The complex numbers having $y = 0$ (i.e., of the form x or $(x, 0)$) are called *real* numbers. These are the points of the x-axis. Those having $x = 0$ (i.e., of the form iy or $(0, y)$) are called *pure imaginaries*. These are the points of the y-axis. For any complex number $x + iy$, the two perpendicular projections on the coordinate axes are x and iy. The real numbers x, y are called the *real* and *imaginary parts* of $x + iy$.

In order for the complex numbers to form a number system, we must define the operations of addition and multiplication on complex numbers. We add two complex numbers by adding separately their real parts and their imaginary parts.

$$(x_1, y_1) + (x_2, y_2) = (x_1 + x_2, y_1 + y_2),$$

or, equivalently,

$$(x_1 + iy_1) + (x_2 + iy_2) = (x_1 + x_2) + i(y_1 + y_2).$$

The geometric picture of addition is that of *vector addition*, where each complex number (x, y) is pictured as a vector from the origin to the point (x, y). The sum of two vectors is just the diagonal of the parallelogram completed on the vectors to be added (see Fig. 35.2).

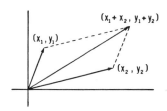

Figure 35.2

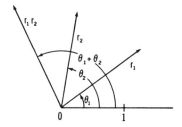

Figure 35.3

Multiplication is more complicated. In terms of coordinates, the product is easily formed by the rule

$$(x_1, y_1) \cdot (x_2, y_2) = (x_1 x_2 - y_1 y_2, x_1 y_2 + x_2 y_1).$$

For example

$$(2, -3) \cdot (-\tfrac{1}{2}, 5) = (14, \tfrac{23}{2}).$$

This rule can be derived by assuming the distributive, associative, and commutative rules for complex numbers, and the extra rule $i^2 = -1$, thus:

$$\begin{aligned}(x_1 + iy_1)(x_2 + iy_2) &= x_1 x_2 + x_1 i y_2 + i y_1 x_2 + i y_1 i y_2 \\ &= x_1 x_2 + i^2 y_1 y_2 + i x_1 y_2 + i x_2 y_1 \\ &= (x_1 x_2 - y_1 y_2) + i(x_1 y_2 + x_2 y_1).\end{aligned}$$

The geometric picture of multiplication is based on angles and lengths of vectors. All angles at the origin are measured from the positive x-axis. Then any vector from the origin is determined by a pair of numbers $[r, \theta]$ where θ is the angle in degrees it makes with the x-axis, and $r \geq 0$ is the length. Two vectors (complex numbers) are multiplied by adding their angles and multiplying their lengths (see Fig. 35.3). For example, $i = [1, 90°]$; hence $i^2 = [1, 180°] = (-1, 0) = -1$.

The derivation of this geometric rule from the algebraic rule is an exercise in trigonometry. In the algebraic rule, we substitute $x_1 = r_1 \cos \theta_1$, $y_1 = r_1 \sin \theta_1$, etc., and then apply addition formulas for sine and cosine.

Much work must be done to justify these definitions. First, one must verify that, for numbers along the x-axis, addition and multiplication are the same as for real numbers. In this way the complex numbers form an enlargement of the real numbers. Next, all of the algebraic laws for real numbers must be proved to hold for complex numbers, e.g., the associative and commutative laws for addition and multiplication and the distributive law. The real number $1 = (1, 0)$ is the unit for complex numbers, i.e., $(1, 0) \cdot (x, y) = (x, y)$. The origin $(0, 0)$ is the zero for both addition and multiplication:

$$(0, 0) + (x, y) = (x, y), \qquad (0, 0) \cdot (x, y) = (0, 0).$$

Finally one must prove that addition and multiplication are continuous operations.

It is customary to abbreviate (x, y) by z, thus $z = x + iy$. Then $z^2 = z \cdot z = (x^2 - y^2, 2xy)$. We define z^n for all integers $n \geq 1$ by the inductive rule $z^n = z \cdot z^{n-1}$. Then $fz = z^n$ defines a mapping $P \to P$ of the set of complex numbers into itself. If $n = 1$, f is just the identity map. If $n = 2$, then f doubles the angle from the x-axis and squares the distance from the origin. Each ray issuing from 0 is mapped onto the ray having twice the angle. A circle of radius r about 0 is mapped onto the circle of radius r^2, and is wrapped around it twice. It is most useful to think of P as a *fan* of rays issuing from 0. Then z^2 wraps the fan twice around itself.

Similarly, $fz = z^n$ multiplies angles by n and raises radii to the n-th power. A circle of radius r about 0 is wrapped n times around the circle of radius r^n about 0. Thus, if C is any circle about 0, the winding number $W(f \restriction C, 0)$ is n for this function $fz = z^n$.

A polynomial f of degree n is defined just as for real numbers. It is a function given by a formula

$$fz = a_n z^n + a_{n-1} z^{n-1} + \cdots + a_2 z^2 + a_1 z + a_0,$$

where $a_0, a_1, \cdots, a_n$ are specified complex numbers, and $a_n \neq 0$. (Keep in mind that each real number is a complex number, and some or all of the a's may be real.) As a function, f defines a mapping $f\colon P \to P$. Its continuity is proved by using the continuity of addition and multiplication of complex numbers.

Exercise

1. Describe geometrically the mapping $f\colon P \to P$ given by each of the

formulas

(a) $fz = z - 4$,   (b) $fz = z + 2i$,
(c) $fz = z + (1 - i)$,   (d) $fz = 2z$,
(e) $fz = -z$,   (f) $fz = iz$,
(g) $fz = (i + 1)z$,   (h) $fz = (i + 1)z - 3i$,
(i) $fz = iz^2 + 2i + 2$,   (j) $fz = 1/z$ $(z \neq 0)$.

36. Every polynomial has a zero

THEOREM 36.1. *Let $n \geq 1$ be an integer, and let f be a polynomial of degree n with complex numbers as coefficients. Then f has at least one zero, that is, there is a complex number α such that $f(\alpha) = 0$.*

Since the coefficient of z^n in f is not zero, we can divide by it and obtain $f/a = g$ or $f = ag$ where g has leading coefficient 1:

$$g(z) = z^n + a_1 z^{n-1} + \cdots + a_{n-1} z + a_n.$$

Since a zero of g is also a zero of f, we need only prove that g has a zero. We shall do this by showing that there is a circle C whose image under the mapping $g: P \to P$ winds about the zero point n times: $W(g \mid C, 0) = n$. Since $n \neq 0$, the main theorem of Part II asserts that there is a point α inside C such that $g\alpha = 0$.

The distance $d(z, 0)$ of a complex number z from 0 is called its absolute value, and it is abbreviated by $|z|$. The circle C will have center at 0 and its radius r is any number larger than the maximum r_0 of the real numbers

$$n|a_1|, \quad (n|a_2|)^{1/2}, \quad \cdots, \quad (n|a_n|)^{1/n}.$$

The direct computation of $W(g \mid C, 0)$ is too difficult, so we shall construct a homotopy of $g \mid C$ into the simpler mapping given by the polynomial z^n. In Section 35 we saw that $W(z^n \mid C, 0) = n$. So if we can show that 0 is not in the image of the homotopy, then the constancy of the winding number implies $W(g \mid C, 0) = n$.

Define the homotopy of $g \mid C$ by the formula

$$g(z, \tau) = z^n[1 + (1 - \tau)h(z)], \quad z \in C, \quad 0 \leq \tau \leq 1,$$

where

$$h(z) = \frac{a_1}{z} + \frac{a_2}{z^2} + \cdots + \frac{a_n}{z^n}.$$

When $\tau = 1$, $g(z, \tau)$ reduces to z^n; and when $\tau = 0$, we find that

§36] EVERY POLYNOMIAL HAS A ZERO 131

$g(z, 0) = gz$. It remains to show that 0 is not in the image of this homotopy, that is, $g(z, \tau) \neq 0$ for all $z \in C$ and all $\tau \in [0, 1]$.

Now $z \in C$ means $|z| = r$. Since $r > (n |a_k|)^{1/k}$ for each $k = 0, 1, \cdots, n - 1$, it follows that $r^k > n |a_k|$, and this implies that, for all $z \in C$,

$$\frac{|a_k|}{|z^k|} = \frac{|a_k|}{r^k} < \frac{1}{n}.$$

Since $h(z)$ has n terms each of absolute value less than $1/n$ on C, it follows that $|h(z)| < 1$ on C. Since also $|1 - \tau| \leq 1$, we find that $|(1 - \tau)h(z)| < 1$. The sum of 1 and a complex number of absolute value less than 1 is never zero. Therefore, for all $z \in C$ and $\tau \in [0, 1]$, $1 + (1 - \tau)h(z)$ is not zero. Since z^n is also not zero for $z \in C$, it follows that the product $z^n[1 + (1 - \tau)h(z)] = g(z, \tau)$ is not zero. This completes the proof.

An algebraist would not be content with only one zero of a polynomial of degree greater than 1. The following theorem gives the complete result.

THEOREM 36.2. *Let f be a polynomial of degree n with (real or) complex numbers as coefficients. Then there are n complex numbers $\alpha_1, \alpha_2, \cdots, \alpha_n$ such that f factors into the product of the n linear factors*

$$f(z) = a_n(z - \alpha_1)(z - \alpha_2) \cdots (z - \alpha_n).$$

If we set $z = \alpha_i$, the i-th factor is zero; hence each α_i is a zero of f. If we set $z = \alpha$, where α is a number different from $\alpha_1, \cdots, \alpha_n$, then each factor is non-zero; hence $f(\alpha) \neq 0$. This shows that $\alpha_1, \alpha_2, \cdots, \alpha_n$ are zeros of f, and they are the only zeros. In case a particular number occurs two or more times in the sequence, it is called a multiple zero, and the number of its occurrences is called its multiplicity.

To prove the theorem we need the following lemma.

LEMMA. *If f is a polynomial of degree n and if α is any complex number, then there is a polynomial g of degree $n - 1$ such that $f(z) = (z - \alpha)g(z) + f(\alpha)$.*

We prove the lemma by dividing f by $z - \alpha$ using the process of long division. Let g denote the quotient and r the remainder, so that

$$\frac{f(z)}{z - \alpha} = g(z) + \frac{r}{z - \alpha}.$$

Multiplying by $z - \alpha$ gives

$$f(z) = (z - \alpha)g(z) + r.$$

To evaluate r, we set $z = \alpha$, and obtain $f(\alpha) = r$.

To prove Theorem 36.2, we apply Theorem 36.1 which says that f has at least one zero, say α_1. Since $f(\alpha_1) = 0$, the lemma states that there is a polynomial g of degree $n - 1$ such that

$$f(z) = (z - \alpha_1)g(z) + f(\alpha_1) = (z - \alpha_1)g(z).$$

The remainder term drops out because $f(\alpha_1) = 0$. If $n - 1 \geq 1$, we may apply Theorem 36.1 to obtain a zero α_2 of g. Then the lemma says that there is a polynomial h of degree $n - 2$ such that

$$g(z) = (z - \alpha_2)h(z).$$

Combining these gives

$$f(z) = (z - \alpha_1)(z - \alpha_2)h(z).$$

If $n \geq 3$, there is a zero α_3 of h, and $h(z) = (z - \alpha_3)k(z)$; whence

$$f(z) = (z - \alpha_1)(z - \alpha_2)(z - \alpha_3)k(z).$$

Each step reduces the degree of the last factor by 1. After n steps, the last factor has degree 0; hence it is a constant c, and

$$f(z) = (z - \alpha_1)(z - \alpha_2) \cdots (z - \alpha_n)c.$$

If we multiply out the right side, the coefficient of z^n is c. Therefore $c = a_n$. This completes the proof.

Historical comment. Theorem 36.1 has been called the fundamental theorem of algebra. Its first rigorous demonstration was given by Gauss in 1797 (see D. E. Smith, *A Source Book of Mathematics*, Dover 1959, page 292). In later years he gave several quite different proofs, but none resembles the one presented above. A direct and fairly simple proof along classical lines can be found in the book *Calculus* by Ford and Ford (McGraw–Hill, 1963), page 263.

Exercises

1. If r_0 is defined as in the proof of Theorem 36.1, and E denotes the set $|z| > r_0$, show that f has no zeros in E.

2. Compute the number r_0 for each of the following polynomials, thus finding a disk about the origin which contains *all* the zeros of the polynomial.
   (a) $z^4 + 3z^3 - 2z + 5$, (b) $2z^7 + iz^3 - 3z$, (c) $z^4 + (2 - i)z^3 + (i + 1)z$.

3. Let $fz = (z - 2)(z + 1)(z - i)4$; compute the coefficients of the polynomial, compute r_0 as above, and check that the disk with radius r_0 contains all the zeros.

4. Find the zeros of each of the following quadratic polynomials and verify that all zeros lie inside the disk with radius r_0.

(a) $3z^2 - 13z - 10$,   (b) $3z^2 - 2z + 1$.

5. In the proof of the theorem, show that it suffices to pick r so that the sum

$$\frac{|a_1|}{r} + \frac{|a_2|}{r^2} + \cdots + \frac{|a_n|}{r^n} < 1.$$

Using this fact, show that all the zeros of $z^3 - z + 5$ lie inside the circle $|z| = 2$.

37. Epilogue: A brief glance at higher dimensional cases

In passing from the one-dimensional case to the two-dimensional, we met with serious difficulties which could only be resolved by the development of a new concept—the winding number $W(\varphi, y)$ of a closed curve φ about a point y. One would expect additional difficulties in passing to the three- and higher dimensional cases. Indeed they do appear, but much that we have done in the two-dimensional case carries over with little change. A brief sketch of a part of what is known about this problem is worthwhile because it embodies some of the best of modern research in mathematics.

In passing from the plane P to a n-dimensional euclidean space R^n, it is natural to replace the disk and its boundary circle by the spherical ball B and its boundary sphere S. Let $f: B \to R^n$ be a mapping, and y a point of R^n not in fS. Then the main theorem states: If $W(f \mid S, y) \neq 0$, then there is at least one $x \in B$ such that $fx = y$. The chief problem lies in defining the number $W(f \mid S, y)$ so that it has all the properties of the winding number when $n = 2$. When $n = 3$ it is better to call $W(f \mid S, y)$ the *enclosing* number. For example, the sphere S should enclose each interior point of B exactly once. This work has been carried out: the number $W(f \mid S, y)$ has been defined precisely for all dimensions n, and has been shown to have the same properties as for $n = 2$. For example, it is unchanged by a homotopy that avoids y.

Once the main theorem has been proved, then the applications discussed in Sections 27–36 for $n = 2$ can be stated and proved for all dimensions with only moderate changes in the notation and language. Let us state a few of these. Since the sphere S encloses once each interior point of the ball B, any mapping $f: B \to R^n$, which leaves fixed all points of S, has the property $fB \supset B$. Next we have that any mapping $B \to B$ has at least one fixed point. Again, if a sphere S in R^n is mapped into R^{n-1}, then some pair of antipodes have the same image point.

The theorem about a field of tangent vectors to a sphere S in R^n

must be modified. When n is even, S does have a continuous field of non-zero tangent vectors (e.g., when $n = 2$, S is a circle). When n is odd, any continuous field of tangent vectors to S has at least one zero.

In jumping to the n-dimensional case, we have skimmed over two questions worthy of much more attention. The first of these is: Can the ball and sphere be replaced in the main theorem by other objects without altering the truth of the conclusion? Naturally, one can always replace them by topologically equivalent objects such as a solid box and its boundary surface. But can we replace them by objects which are topologically different without altering essentially the conclusion of the main theorem? The answer is "yes" for dimensions greater than 2. For example, in R^3 let T be a torus and D its interior (e.g., D is a solid ring and T is its boundary surface). One can define an enclosing number $W(f \mid T, y)$ so that T encloses once each inner point of D, and encloses each point of the complement of D zero times. Other examples in R^3 are provided by the multiply-connected surfaces and their solid interiors. Figure 37.1 illustrates a triple doughnut and its boundary.

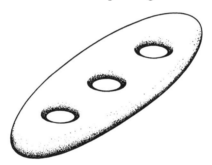

Figure 37.1

The second question we have glossed over is: Why do we consider only mappings of n-dimensional sets into n-dimensional sets? Can our main theorem be generalized so as to allow mappings of a k-dimensional set into R^n? We shall indicate briefly how something can be done in this direction in the case $k = 2$ and $n = 3$. Let $f: D \to R^3$ be a mapping of a disk into space. Let $\varphi: [a, b] \to R^3$ be a closed curve in R^3 which does not intersect fC. Assigned to the two curves $f \mid C$ and φ is an integer $W(f \mid C, \varphi)$ called their *linking number*. The five examples in Fig. 37.2, ordered from left to right, have linking numbers 0, 1, 2, 4, and 0. The new version of the main theorem reads as follows: *If the linking number $W(f \mid C, \varphi)$ is not zero, then fD intersects the closed curve φ in at least one point*. The reader should check this statement against the five examples in Fig. 37.2. In the first and fifth examples, one is able to picture a surface whose boundary is the lower closed curve, and which misses the upper closed curve. This surface might be fD. In the other three examples, no matter how we insert a surface fD whose boundary is the lower curve, this surface intersects the upper curve. (In examples 3 and 4, one can picture a twisted strip, i.e. a Möbius band, which does not meet the upper curve and whose periphery is the lower curve; however it cannot be an fD because it is one-sided.)

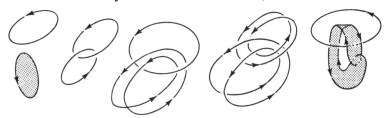

Figure 37.2

Notice that the winding number $W(f \mid C, y)$ in the plane involves a closed curve and a point, but, on passing to maps of a disk into R^3, the point y is replaced by a closed curve φ, and the winding number becomes a linking number. If we were to formulate an analog of our main theorem for mappings of a disk in R^4, we would need the concept of "linking number" $W(f \mid C, \varphi)$ of a closed curve $f \mid C$ and a closed surface φ in R^4. By a closed surface in R^4 we mean a continuous mapping into R^4 of a sphere or a torus, or any one of the multiply-connected surfaces.

Thus a point, a closed curve, and a closed surface are examples in dimensions 0, 1, and 2 of a concept defined for all dimensions called a *cycle*. The things of which they are the boundaries (e.g. intervals, disks, balls, etc.) are called *chains*. Cycles, chains, their homologies and homotopies, and their intersections and linkings make up the main fare of the fascinating subject of homology theory which is a major part of topology.

We have shown how some of the simpler ideas of topology can be used to prove theorems which are, at the same time, intuitively satisfying yet subtle. They are existence theorems. Higher dimensional generalizations of these theorems can be formulated and proved using the concepts of homology theory.

A reader who desires to pursue the development of the ideas presented in this monograph can consult the following books. The one by Hall and Spencer provides a continuation of the material of Part I on point-set topology. The other two continue the ideas of Part II on algebraic topology.

D. W. Hall and G. L. Spencer, *Elementary Topology*, John Wiley and Sons, New York, 1955.

J. G. Hocking and G. S. Young, *Topology*, Addison–Wesley, Reading, Mass., 1961.

P. J. Hilton and S. Wylie, *Homology Theory*, Cambridge University Press, 1960.

Solutions for Exercises

Section 1

1. Minimum $f(3) = 1$, maximum $f(1) = 5$, since $f(x) = 5 - (x-1)^2$. No solution if $y < 1$ or if $y > 5$. One solution if $1 \leq y < 4$ or if $y = 5$. Two solutions if $4 \leq y < 5$.

2. Since $x^3 - 5$ takes on all values between -4 and $+3$ as x varies from 1 to 2, for some x between 1 and 2 it takes on the value 0. Since $x^3 - 5 = 0$ means $x^3 = 5$, such an x must be $\sqrt[3]{5}$.

3. The value of this function at $x = 3$ is negative, at $x = 4$ is positive, so it has a zero between 3 and 4.

4. Minimum $f(5) = \frac{1}{5}$. There is no maximum because $f(1/n) = n$ for $n = 1, 2, 3, \cdots$ are values of f. Note that $f(0)$ is not defined.

5. In this case, every value of f is 3 (its graph is a horizontal line segment), so $m = M = 3$.

6. Minimum $f(0) = 0$. There is no maximum because 5 is not in the interval $[0, 5)$.

Section 2

1. To prove the equality of two sets, one must show that an element of the first set is an element of the second set and vice versa. For the first part, suppose $x \in (A \cup B) \cap C$. This means that x is in A or B, and also x is in C. So we must consider two cases.
   Case 1: $x \in A$ and $x \in C$. It follows that $x \in A \cap C$, and this implies that $x \in (A \cap C) \cup (B \cap C)$.

Case 2: $x \in B$ and $x \in C$. It follows that $x \in B \cap C$, and this implies that $x \in (A \cap C) \cup (B \cap C)$.

For the second part, suppose $x \in (A \cap C) \cup (B \cap C)$. This means that $x \in A \cap C$ or $x \in B \cap C$.
Case 1: $x \in A \cap C$. It follows that $x \in A$ and $x \in C$. This implies that $x \in (A \cup B)$ and $x \in C$; hence $x \in (A \cup B) \cap C$.
Case 2: $x \in B \cap C$. It follows that $x \in B$ and $x \in C$. This implies that $x \in (A \cup B)$ and $x \in C$.
In either case, $x \in (A \cup B) \cap C$. (See Fig. S1.)

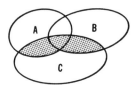
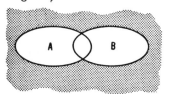

Figure S1   Figure S2

2. (See Fig. S2.)

3. Since B and its complement have no points in common, any subset of B and the complement of B have no points in common.

4. (a) Following the pattern described in the answer to Problem 1, let $y \in f(A \cup B)$. This means that there is an $x \in A \cup B$ such that $fx = y$. In case $x \in A$, we have $y \in fA$, and in case $x \in B$, we have $y \in fB$. Thus, in either case $y \in fA \cup fB$. For the converse argument, let $y \in fA \cup fB$. In case $y \in fA$, then there is an $x \in A$ such that $fx = y$; and since $x \in A \cup B$, this means $y \in f(A \cup B)$. In case $y \in fB$, $y = fx$ for some $x \in B$; and since $x \in A \cup B$, this means $y \in f(A \cup B)$. So, in either case, $y \in f(A \cup B)$.

(b) Let $y \in f(A \cap B)$, then $y = fx$ for some $x \in A \cap B$. Since $x \in A$, we have $y \in fA$; and since $x \in B$, we have $y \in fB$. Therefore $y \in fA \cap fB$. (The converse argument breaks down. Suppose X consists of two points A and B, and Y has just one point y; then $fA = y = fB$, $f(A \cap B) = \emptyset$, and $fA \cap fB$ consists of y.)

5. (a) A circle with center at S.
(b) A line through S.
(c) An arc of a circle (a plane through N and the segment cuts X in a circle).
(d) Twice as large.
(e) All points of a great semicircle from N to S but omitting N.

6. gf: $y_1 = -x_1 - 3,$ $y_2 = x_2 - 4$
fg: $y_1 = -x_1 + 3,$ $y_2 = x_2 - 4$
f^{-1}: $y_1 = x_1 - 3,$ $y_2 = x_2 + 4$
g^{-1}: $y_1 = -x_1,$ $y_2 = x_2$
$(gf)^{-1}$: $y_1 = -x_1 - 3,$ $y_2 = x_2 + 4$
$f^{-1}g^{-1}$: $y_1 = -x_1 - 3,$ $y_2 = x_2 + 4$

$g^{-1}f^{-1}$: $y_1 = -x_1 + 3$, $y_2 = x_2 + 4$
No; $g^{-1}f^{-1} \neq (gf)^{-1}$.

7. (b) $f^{-1}Y = X$; $f^{-1}\emptyset = \emptyset$.   (c) $f^{-1}A \subset f^{-1}B$.

8. $x \in (gf)^{-1}C$ means that $gfx \in C$; and this means that $fx \in g^{-1}C$, which in turn means that $x \in f^{-1}(g^{-1}C)$; so $x \in (gf)^{-1}C$ means that $x \in f^{-1}(g^{-1}C)$. (We use the word "means" to mean that the statements are equivalent, i.e. each implies the other.) For the second part, suppose $x \neq x'$. Since f is 1—1, we have $fx \neq fx'$, and since g is 1—1, we have $gfx \neq gfx'$. Now let $z \in Z$. Since $gY = Z$, there is a $y \in Y$ such that $gy = z$; and since $fX = Y$, there is an $x \in X$ such that $fx = y$; hence $gfx = z$ or $x = (gf)^{-1}z$. By definition, $y = g^{-1}z$ and $x = f^{-1}y$; hence $x = f^{-1}g^{-1}z$. Therefore $(gf)^{-1} = f^{-1}g^{-1}$.

Section 3

1. When x, y, z are in line and y is between x and z.

2. $r' = r - d(x, x')$.

3. Because $N(fx, \epsilon) \supset N(fx, \frac{1}{2})$ for all $\epsilon \geq \frac{1}{2}$.

4. Because $N(x, \delta') \subset N(x, \delta)$ for all $\delta' < \delta$.

5. Discontinuous at each point of C, continuous everywhere else. For $\epsilon \leq \sqrt{2}$, there is no corresponding δ.

6. Because it does not increase any distance:
$$d(x, x') \geq d(fx, fx') \text{ for all } x, x'.$$

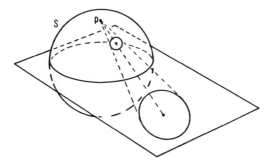

Figure S3

7. Contracts distances for any pair of points in the lower hemisphere. f^{-1} sends $y \in R^2$ into the intersection of the ray py with S. See Fig. S3.

140   FIRST CONCEPTS OF TOPOLOGY

The same picture shows that the inverse is continuous. The image of $N(p, r, S) - p$ is the exterior of a circle with center at the origin. If fp were defined and f were continuous at p, then any $N(fp, \epsilon)$ would have to contain the entire exterior of some circle; but this is impossible.

8. Let $x \in [a, b]$ and $\epsilon > 0$ be fixed. Since f is continuous at x, there is a $\delta_1 > 0$ such that $|fx' - fx| < \epsilon/2$ for all $x' \in N(x, \delta_1)$. Since g is continuous, there is a $\delta_2 > 0$ such that $|gx' - gx| < \epsilon/2$ for all $x' \in N(x, \delta_2)$. Let δ be the smaller of δ_1, δ_2. Then $x' \in N(x, \delta)$ implies

$$|fx' - fx| + |gx' - gx| < \epsilon/2 + \epsilon/2 = \epsilon.$$

Apply now the inequality of the hint.

9. $\delta = d \sin 2\theta = 2d \sin \theta \cos \theta$, where $\sin \theta = \epsilon/2$. Since

$$\cos \theta = \sqrt{1 - \sin^2 \theta},$$

direct substitution gives $\delta = \epsilon d \sqrt{1 - \epsilon^2/4}$.

Section 4

1. A single point $x \in X$ is a closed set of X for any space X. Since a finite union of closed sets is closed, any finite set A of a space X, being the finite union of its single points, must be closed. If X is also finite, then $X - A$ is finite; hence $X - A$ is closed, so A is open.

2. One solution is $V = U \cup (R^2 - L)$.

3. $x^2 + y^2 < 1$.

4. A set of a single point; also R^2.

5. Let X consist of two points x_1, x_2, $A = \{x_1\}$, and $B = \{x_2\}$. A and B are complements. $A \cup B = X$; $X - (A \cup B) = \emptyset$;

$$(X - A) \cup (X - B) = B \cup A = X.$$

6. Let $X = R$, $A = (0, 2]$, $B = [1, 3)$; then $A \cup B$ is open in R.

7. If U is open in X, then by Theorem 4.4 there is an open set W of R^n such that $U = X \cap W$. Let $V = Y \cap W$. Then V is open in Y, and $U = X \cap V$ because $X \cap Y = X$.

8. Intersection of these intervals is the single point 0, and a single point of R is never an open set of R.

9. Let $X = \{x_1\}$, and $Y = \{y_1, y_2\}$ such that $fx_1 = y_1$. $A = \emptyset \subset X$,

$fA = \emptyset$, $\quad Y - fA = Y$. $\quad f(X - A) = fX = y_1$. $\quad$ So $Y - fA \neq f(X - A)$.

10. $x \in X - f^{-1}A$ means that $x \in X$ and $x \notin f^{-1}A$. $x \in X$ means $fx \in Y$, and $x \notin f^{-1}A$ means $fx \notin A$. So $fx \in Y - A$, and this means that $x \in f^{-1}(Y - A)$.

11. Let A be a subset of X such that A is open in X; then, for each $x \in A$, there is some neighborhood of x in X which lies in A. Each neighborhood is open; take the union of all such neighborhoods.

Section 5

1. If $\sqrt{3}$ were rational, we could write $\sqrt{3} = a/b$, where a, b are integers with no common factor. Square both sides, and multiply by b^2 to obtain $a^2 = 3b^2$ which shows that a^2 is divisible by 3. If a were not divisible by 3, it would be of one of the forms $a = 3k + 1$ or $a = 3k + 2$ for some integer k. In the first case

$$a^2 = 9k^2 + 6k + 1 = 3(3k^2 + 2k) + 1 = 3k' + 1;$$

in the second case

$$a^2 = 9k^2 + 12k + 4 = 3k'' + 1;$$

in either case a^2 is not divisible by 3. But $a^2 = 3b^2$ is divisible by 3, so a must be divisible by 3, and $a = 3k$ for some integer k. Then $3b^2 = (3k)^2 = 9k^2$; whence $b^2 = 3k^2$; this says that b^2 is divisible by 3 from which it follows that b is divisible by 3. This is a contradiction because a, b were assumed to have no common factor.

2. Let $2n^2 = m^2$; since n is an integer, its prime factorization contains either an odd number of 2's or an even number of 2's. In either case, n^2 has an even number of 2's, and $2n^2$ has an odd number of 2's. But m^2 has an even number of 2's as factors; so if m and n are integers, $2n^2 \neq m^2$; that is, $2n^2 = m^2$ has no solution in integers m, n.

3. Let $2n^3 = m^3$; if the factor 2 occurs k times in m, it occurs $3k$ times in m^3. If 2 occurs k' times in n, then it occurs $3k'$ times in n^3 and $3k' + 1$ times in $2n^3$. Since the number of any prime factor in a complete factorization is unique, $2n^3 = m^3$ has a solution in integers only if there are integers k, k' such that $3k' + 1 = 3k$. This is impossible since $3k' + 1$ is not divisible by 3 but $3k$ is divisible by 3.

4. If there were rational solutions $x = m/n$ and $y = p/q$, where m, n, p, q are integers, then

$$\frac{p^2}{q^2} = \frac{2m^2}{n^2}, \quad \text{so} \quad (np)^2 = 2(mq)^2, \quad \text{and} \quad \sqrt{2} = +\frac{np}{mq}.$$

This says that $\sqrt{2}$ is rational, contradicting what we have proved in Section 5.

5. If there are integers k, m, n such that

$$\left(\frac{k + m\sqrt{2}}{n}\right)^3 = 2, \quad \text{and} \quad m \neq 0,$$

then cubing gives $k^3 + 3k^2m\sqrt{2} + 3km^2 2 + 2m^3\sqrt{2} = 2n^3$, from which it follows that

$$\sqrt{2} = \frac{2n^3 - k^3 - 6km^2}{3k^2m + 2m^3}$$

is a rational number; contradiction. The case of $m = 0$ is that of Exercise 3.

6. The partial sums of this series are

$$0, \quad 1, \quad .9, \quad .91, \quad .909, \quad .9091, \quad .90909, \quad \cdots$$

and the intervals from each sum to the next form a sequence of contracting intervals

$$[0, 1] \supset [.9, 1] \supset [.9, .91] \supset [.909, .91] \supset \cdots .$$

These contract regularly. Their intersection is the number whose decimal expansion is $0.909090\cdots$ (repeating on the digits 90). It equals $10/11$ and represents the sum of the infinite series.

7. Similar to Exercise 6:

$$[0, 1] \supset [\tfrac{1}{2}, 1] \supset [\tfrac{1}{2}, \tfrac{3}{4}] \supset [\tfrac{5}{8}, \tfrac{3}{4}] \supset \cdots .$$

The sum is $\tfrac{2}{3}$.

8. First trial shows that 3.41 divides 12.27 at least 3 times but not 4, so first interval is $[3, 4]$. Second trial shows that 3.41 divides 12.27 at least 3.5 times but not 3.6, so second interval is $[3.5, 3.6]$. The next is $(3.59, 3.60]$, and then $[3.598, 3.599]$, $\cdots$.

9. $[1, 2] \supset [1.5, 1.6] \supset [1.58, 1.59] \supset \cdots .$

10. If s is a positive number less than $b - a$, then all the integral multiples of s are evenly spaced along the line with each two neighbors being a distance s apart. Since a and b are farther apart than s, the interval (a, b) must contain a multiple of s. To construct a rational in (a, b), choose $s = 1/n$ where n is an integer such that $n > 1/(b - a)$. It follows that $s = 1/n < b - a$, and that some multiple $ms = m/n$, a rational number, lies in (a, b). To obtain an irrational number contained in (a, b), let $s = \sqrt{2}/n < b - a$. Q is not closed since, for each $x \in R - Q$ and each $r > 0$, the interval $N(x, r)$ contains rational numbers. Neither is it open because each open interval contains irrational numbers.

11. Since $a < b$ for every $a \in A$ and $b \in B$, pick any $I_0 = [a_0, b_0]$ where $a_0 \in A$ and $b_0 \in B$. Let $c_0 = \frac{1}{2}(a_0 + b_0)$. Then $a_0 < c_0 < b_0$. Either $c_0 \in A$ or $c_0 \in B$. If $c_0 \in A$, let $I_1 = [a_1, b_1]$, where $a_1 = c_0$ and $b_1 = b_0$. If $c_0 \in B$, let $a_1 = a_0$ and $b_1 = c_0$. Continue in this way to build a contracting sequence so that, for each $n \geq 1$, $I_n = [a_n, b_n]$ is a half of I_{n-1}, and $a_n \in A$, $b_n \in B$. By the completeness theorem, the intervals I_n have a point c in common. By construction, if $c \in A$, all preceding a's are less than c, and c is the largest number of A; if $c \in B$, c is the smallest number of B. See proof of theorem on page 34.

Section 6

1. If X is bounded it is contained in some sufficiently large ball B. If $A \subset X$ then A is contained in B, hence any subset A of X is also bounded.

2. If X, Y are two bounded sets then $X \subset N(x_0, r)$, $Y \subset N(y_0, s)$ for some r, s. Let $t = d(x_0, y_0)$; then

$$N(x_0, r + s + t) \supset N(x_0, r) \cup N(y_0, s) \supset X \cup Y.$$

3. For example, each integer is a bounded set; but the infinite sequence $1, 2, 3, \cdots, n, \cdots$ is unbounded.

4. $U_k = \left[a, b - \dfrac{b-a}{2^k}\right)$; also $U_k = \left[a, b - \dfrac{b-a}{k}\right)$, etc.

5. In the first case, U_k consists of all $x \in X$ such that $1/k < d(x_0, x) \leq 1$; that is, U_k is an annulus containing its outer circle but not its inner one. In the second case, U_k consists of all points $x \in X$ such that

$$d(x, y_0) > 1/k \,;$$

that is, U_k is a crescent shaped region.

6. The least number of such intervals is 11; for example, let $s = 1/100$ and set $U_1 = (-s, 1 - s)$, $U_2 = (1 - 2s, 2 - 2s)$, $\cdots$, $U_{11} = (10 - 11s, 11 - 11s)$. This is a subset of all intervals of R of length 1, and since this subset covers X, the set of all such intervals is a covering of X.

7. If y is in the exterior of C_1, then the line segment from x_0 to y meets C_1 in a point c such that U_c contains y. C_r is closed and bounded, hence compact. See Fig. S4. The open annulus cannot be covered by a finite subcollection of C because, as r gets nearer to 1, the number of half-planes required to cover C_r increases without bound.

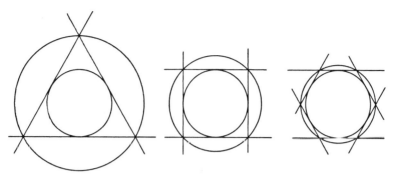

Figure S4

8. The union of a finite covering of X and a finite covering of Y is a finite covering of $X \cup Y$.

9. Take the integers as one-element sets; or consider R as the union of the closed intervals $[-n, n]$, $n = 1, 2, 3, \cdots$.

10. For any point y of $Y - X$, the exterior of $N(y, r)$ for each $r > 0$ is open, and the collection of these for $r > 0$ covers X. Choose a finite covering. Take the smallest r in this covering. Then $N(y, r) \cap X = \emptyset$. This proves that $Y - X$ is open, hence X is closed.

11. Let a and b be irrationals (for example, let $a = -\sqrt{2}$ and $b = \sqrt{2}$); then $X = [a, b] \cap Q$ is closed in Q since $[a, b]$ is closed in R, and X contains neither a greatest nor a smallest rational number.

12. $fx = nx$; $f: [-1, 1] \to [-n, n]$. No; because I is compact so fI is also compact, but R is not compact. The function $fx = x/(1 - x^2)$ maps $(-1, 1)$ onto R; so does $fx = \tan \frac{1}{2}\pi x$. There are many such functions: any continuous increasing function with vertical asymptotes at $x = -1$ and $x = +1$ fulfills the requirement.

13. First, let X be compact and C a covering of X by open sets of R^m. The collection C' of intersections $V' = V \cap X$ for all $V \in C$ is a covering of X by open sets of X. As X is compact, C' contains a finite covering D'. The V's in C corresponding to V' in D' form a finite covering $D \subset C$. Next, let X have the property that each covering of X by open sets of R^m contains a finite covering, and let C be a covering of X by open sets of X. If $y \in R^n - X$, the exteriors of neighborhoods of y form a covering of X by open sets of R^n; since it contains a finite covering, y has a neighborhood not meeting X; hence X is closed. For each $V \in C$, let $V' = V \cup (R^m - X)$. The collection of these for all $V \in C$ is a covering C' of X by open sets of R^m. A finite covering in C' corresponds to one in C.

Section 7

1. (a) The first is connected; the second is not and separates into the two different circular arcs connecting the two deleted points which, of course, do not belong to the separation.

   (b) The first is connected; the second separates into two subarcs.

   (c) The first is not connected, for if A is the set consisting of any point, and B the set consisting of the other points, A, B is a separation. The other two sets are connected.

   (d) (i) Connected (section of rubber tubing);
   (ii) Connected (with cut along Q, an inner tube can roll out into an annulus);
   (iii) Connected (section of rubber tubing sliced lengthwise rolls out into rectangle);
   (iv) Disconnected (patch of inner tube severed from remaining portion);
   (v) Disconnected (inner tube cut into two sections);
   (vi) Disconnected (ring collar sliced from tube);
   (vii) Connected (deletion of two P's from the solid figure is like making two scratch marks on the surface; any two points may be connected via the interior).

   (e) Not connected; A and B are the two circles. $A \cap B = \emptyset$, so the intersection is connected.

   (f) (i) Connected;   (ii) Connected;   (iii) Not connected;
   (iv) Connected;   (v) Connected.

2. If D is star-shaped about the point p, then each $x \in D$ lies on the line segment px in D, and this segment is connected. Any two points x, y of D lie on the broken line $xp \cup py$ in D which is also connected since it is the union of two connected sets with a common point.

3. In all cases, two points outside, inside, or on the surface can be connected by an arc of a circle.

4. Let the domain X consist of two points and the range Y consist of one point.

5. For a tangent line, the projection f shrinks distances, and so is continuous. For a chord, let g be projection from the center onto a segment of the parallel tangent line. Then g is a similarity, hence the composition fg is continuous. A chord is a segment, hence is connected, and hence also the arc is connected, being a continuous image of a connected set.

6. Two semicircles whose intersection consists of the two endpoints.

7. The set of points with x rational forms a family of lines parallel to the

y-axis; points with y rational form a family of lines parallel to the x-axis. The union of these forms a grid. Any two points of the grid are connected by a polygonal path having at most three segments. Points having *exactly* one rational coordinate are separated, for example, by the line $y = x$ into those points above the line and those below. Points having two rational coordinates are separated, for example, by the line $x = \sqrt{2}$.

8. A circle with center $(0, 0)$ and radius r', where r' is irrational, separates X.

9. Let X denote the set, and assume $X \neq \emptyset$. Let x_0 be a point of X. Divide $R - X$ into the set A of those numbers less than x_0 and the set B of those greater than x_0. Since X contains all points between any two of its points (see the proof of Theorem 7.6), it follows that each number of A precedes each of X, and each of X precedes each of B. If A and B are empty, we have $X = R$. As case 1, suppose $A \neq \emptyset$ and $B = \emptyset$. Apply Exercise 11, Section 5, to obtain a number a which is either the largest in A or the smallest in X; if $a \in A$, then X is the open half-line of numbers x such that $a < x < \infty$, and if $a \in X$, then X is the closed half-line $a \leq x < \infty$. Case 2, when $A = \emptyset$ and $B \neq \emptyset$, is similar. In case 3, when $A \neq \emptyset$ and $B \neq \emptyset$, we apply Exercise 11, Section 5 to the cut composed of A and $X \cup B$ to obtain a number a which is the largest of A or the smallest of X. Apply it also to the cut $A \cup X$ and B to obtain a number b which is either the smallest in B or the largest in X. If $a = b$, then X is the single point x_0. If $a < b$, then X consists of the open interval (a, b) with none, one, or both endpoints.

10. Since the intervals are closed, their intersection X is also closed. Suppose X has more than one point. If x and y are any two points of X, the interval $[x, y]$ lies in each interval of the sequence and therefore in X; hence X is connected. Since X is closed and bounded, it is compact, and since X is compact and connected, it is a closed interval.

11. Let $I_0 = [a_0, b_0]$, where $a_0 \in A$ and $b_0 \in B$. If the midpoint m of I_0 is in A, set $I_1 = [m, b_0]$; otherwise set $I_1 = [a_0, m]$. Construct in this way a contracting sequence such that, for each n, $I_n = [a_n, b_n]$ is a half of I_{n-1}, $a_n \in A$ and $b_n \in B$. Let c be a point common to all the I_n. Each neighborhood of c contains I_n for sufficiently large n, hence it contains points of both A and B. So, if $c \in A$, then A is not open, and, if $c \in B$, then B is not open. This contradicts the assumption that A, B is a separation because $c \in I_n \subset I = A \cup B$.

Section 8

1. (a) See answer to Exercise 12, Section 6.

    (b) Restrict an answer to (a) to $[0, 1)$.

(c) Using polar coordinates (r, θ) in the plane Y with origin at the center of X, define $f: X \to Y$ by $f(r, \theta) = (r/(b^2 - r^2), \theta)$ where b is the radius of X. In words, f maps each diameter of X topologically onto the line containing the diameter by a mapping of the type in part (a).

2. By stereographic projection, the circle with one point deleted is topologically equivalent to a line; by 1(a), the line is topologically equivalent to an open interval. Alternately, by measuring arc length from a base point, the deleted circle is mapped topologically onto an open segment whose length is the circumference.

3. For example, the set of all whole numbers; the set of all positive rationals; the set of all positive irrationals; the set $\{0, \frac{1}{1}, \frac{1}{2}, \frac{1}{3}, \cdots\}$; etc.

4. (a) Yes; (b) No; in each neighborhood of 0 there are points not connected with 0; (c) Yes; (d) No.

5. (a) For example, the set of all integers; a line; a closed half-line.
   (b) Let X be closed in R^m, and $x \in X$. For any $r > 0$, $N(x, r)$ lies in the set B_r of all points x' such that $d(x, x') \leq r$. Since B_r is closed and bounded, it is compact. Since X is closed, $X \cap B_r$ is compact. It contains $N(x, r, X)$.

6. (a) Not topological; a line is unbounded and is topologically equivalent to an open interval which is bounded.
   (b) Topological.
   (c) Not topological; any two segments of same or different lengths are topologically equivalent.
   (d) Topological.
   (e) Not topological; for example, a square is topologically equivalent to any quadrilateral, convex or concave.
   (f) Topological.
   (g) Topological.

7. X, Y are topologically equivalent, so there is a continuous 1–1 function f such that $fX = Y$. Let C be an open covering of Y; for each $U \in C$, let $U' = f^{-1}U$, and let C' be the collection of these sets U'. Since f is continuous, each U' is open. Since C covers fX, it follows that C' covers X. Since X is compact and C' is a covering of X by open sets, C' contains a finite covering, say $U_1', U_2', \cdots, U_k'$. Then the corresponding sets $U_1, U_2, \cdots, U_k$ in C cover Y because $Y = fX$ and each $U_i = fU_i'$. Therefore Y is compact.

8. (a), (d), (e), (g).

Section 9

1. $1/\sqrt{2}$.

2. (a) The graph is an inverted parabola passing through $(0, 0)$ and $(1, 0)$; its highest point is at $(1/2, 1)$.
   (b) No, for example, $f0 = f1 = 0$.
   (c) $f0 = 0$ and $f(3/4) = 3/4$.

3. (a) The graph is a parabola passing through $(0, 1)$ and $(1, 1)$; its lowest point is at $(1/2, 3/4)$.
   (b) No, for example, $f0 = f1 = 1$, also $f^{-1}0$ is empty.
   (c) $f1 = 1$.

4. We show that every set Y topologically equivalent to X necessarily has the same property. Let $h: X \to Y$ be a homeomorphism, and let $f: Y \to Y$ be continuous. Then $h^{-1}fh: X \to X$ is continuous. Let $x \in X$ be a fixed point of $h^{-1}fh$ so that $h^{-1}fhx = x$. This implies $fhx = hx$, hence $hx \in Y$ is a fixed point of f.

5. x^2, x^3, $\sqrt{x}$, or x^m for any positive m different from 1.

6. Each of the answers to Exercise 5.

7. Suppose f maps $[0, 1)$ onto $[0, 1)$. Then there is a $b \in [0, 1)$ such that $fb = 0$. If $b = 0$, then 0 is fixed; so suppose $b > 0$. Then $fx - x$ is positive for $x = 0$ and negative for $x = b$. By the main theorem $fx - x = 0$ for some $x \in [0, b]$.

Section 10

1. (a) The tangents from p to C divide L into three subsets; namely, the two points of intersection a, b of the tangents with L, the points between a and b, and the points not in the segment ab. For each point of L between a and b, there are two inverse images; for each of a and b, there is one inverse image; for each point of L not in the segment, there is no inverse image.

   (b) The points of intersection of the diametrical line through p with C.

2. Each point of L has just one inverse image. The projection does not map p' into L; and it cannot be defined to do so continuously because C is compact and L is not.

3. Let $f: D \to D'$ be any mapping and compose f with the reflection $g: D' \to D$. Then $gf: D \to D$ has a fixed point x, and $gfx = x$ means that fx is the reflection of x.

4. Compose f with the antipodal mapping $g: D' \to D$. Then $gf: D \to D$ has a fixed point x, and $gfx = x$ means that fx is antipodal to x.

SOLUTIONS   149

5. Let $f: C \to C$ wrap C twice around itself so as to double the arc length from a fixed reference point (e.g., if $\{r, \theta\}$ denote *polar* coordinates in the plane with pole at the center of C, let $f\{r, \theta\} = \{r, 2\theta\}$). Then f maps each diametrical pair into one point. Now compose f with any non-constant mapping $g: C \to L$, e.g., a projection.

Section 11

1. The cut through the line of centers solves the problem; this applies equally to any pair of pancakes with the property that each is symmetric with respect to some point, the center.

2. Only if the polygons have an even number of sides; if the polygon has an odd number of sides, the method works only if a vertex lies on the line of centers.

3. There are infinitely many ways: any perpendicular pair through the center.

4. A quarter turn applied to any solution gives the same solution.

5. Enclose both pancakes within a sufficiently large circle C having the center z of the circular pancake as center. Let r be the radius of C. For each $x \in C$, let P_x be the area of the part of the irregular pancake that is to the left of xz (oriented from x through z), and let Q_x be the area of the part to the right of xz. Define fx to be the difference $P_x - Q_x$. At the diametrical point x' of x, the sides are reversed; hence $fx' = -fx$. Since f is continuous and changes sign as we traverse a semicircle, it is somewhere zero.

6. With a semicircular blade, it would not be true that $x_A' = x_A$ (a semicircle is not transformed onto itself by a 180° rotation about its midpoint). The argument holds for any curved blade that is invariant under a 180° rotation, e.g. an S-shaped blade.

Section 12

1. Observe first that for any x and x' we have

$$|x^2 - x'^2| = |x + x'||x - x'| \leq (|x| + |x'|)|x - x'|.$$

If x and $\epsilon > 0$ are given, take δ to be the smaller of 1 and

$$\epsilon/(2|x| + 1).$$

If x' is such that $|x' - x| < \delta$, then $\delta \leq 1$ implies $|x'| < |x| + 1$ and hence $|x| + |x'| < 2|x| + 1$; and $\delta \leq \epsilon/(2|x| + 1)$ implies $|x^2 - x'^2| \leq (|x| + |x'|)|x - x'| < (2|x| + 1)|x - x'| \leq \epsilon$.

2. Let x and $\epsilon > 0$ be given. Since g is continuous at x, there is a $\delta_g > 0$ such that for all $x' \in N(x, \delta_g)$

$$|gx - gx'| < \frac{\epsilon}{2(|fx| + 1)} \quad \text{and} \quad |gx - gx'| < 1.$$

These imply

$$|fx| |gx - gx'| < \epsilon/2 \quad \text{and} \quad |gx'| < |gx| + 1.$$

Since f is continuous at x, there is a $\delta_f > 0$ such that for all $x' \in N(x, \delta_f)$

$$|fx - fx'| < \frac{\epsilon}{2(|gx| + 1)}.$$

Then

$$|gx'| |fx - fx'| < (|gx| + 1) |fx - fx'| < \epsilon/2.$$

Now take δ to be the smaller of δ_g and δ_f. Then for all $x' \in N(x, \delta)$
$$|fx| |gx - gx'| + |gx'| |fx - fx'| < \epsilon/2 + \epsilon/2 = \epsilon.$$

3. The constant term.

4. By the criterion, $b \geq 6$. The polynomial factors into
$$fx = x(x - 3)(x + 1),$$
so $fx > 0$ for $x > 3$, and $fx < 0$ for $x < -1$.

5. $b = 25$.

Section 13

1. (a) A horizontal strip whose width is the diameter of Q.

   (b) A horizontal line.

   (c) A vertical line segment cutting across the strip.

   (d) The sloping line transforms into a helix on Q, and projects into a curve in the shape of a wave that oscillates back and forth across the strip.

   (e) If the point p is not on the strip, then $f^{-1}p = \emptyset$. If p is on the edge of the strip, $f^{-1}p$ is a sequence of points evenly spaced along a vertical line; the distance between adjacent points is $2\pi r$, where r is the radius of the cylinder. If p is inside the strip, $f^{-1}p$ is again a sequence of points along a vertical line, and the spacing of alternate pairs is $2\pi r$.

2. (a) The image fP consists of all points of P whose distance from the origin is at most the radius of S.

SOLUTIONS 151

(b) Under stereographic projection, the image of a line L is the circle C in which the plane through p and L meets S, with p deleted from C. If C lies on the hemisphere of p, then $C - p$ projects into an ellipse through the origin with the origin omitted. If C meets both hemispheres, it projects into a pair of elliptical arcs each connecting two points on the edge of fP, and one of the arcs passes through the origin.

(c) The inverse image $f^{-1}q$ is q itself when q is the origin or when q is on the edge of fP. For other points of fP, $f^{-1}q$ is a pair of points.

3. (a) A line through z.

   (b) A circle of radius r.

   (c) A spiral whose distance from z increases at a constant rate.

   (d) See Fig. S5.

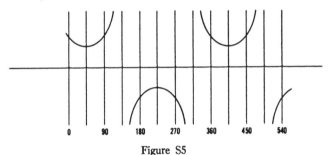

Figure S5

Section 14

1. If D is the disk and C is its bounding circle, let S be a semicircle of C. Define $f \mid S$ to be the identity map, and, on $D - S$, let f map each line segment perpendicular to the diameter of S into itself by a contraction to half its size towards its endpoint on S. Then f is a 1 to 1 mapping of the disk onto the semicircular region T. Any other configuration A that is topologically equivalent to the disk has a 1 to 1 mapping $g: A \leftrightarrow D$; the composition with f gives a 1 to 1 continuous function $fg: A \leftrightarrow T$.

2. Map each radius zx rigidly onto the radius zgx.

3. The common arc is topologically equivalent to a segment, hence, to a diameter of a disk. From Exercise 1, this equivalence extends to a homeomorphism from A to one of the semicircular regions of this disk, and it also extends to a homeomorphism from B to the other semicircular region. So $A \cup B$ is homeomorphic to the disk.

4. h and c

152   FIRST CONCEPTS OF TOPOLOGY

5. (a) No.
   (b) Any two of the three cuts shown in Fig. 14.3 will produce a homeomorph of a disk.
   (c) Three cuts.

Section 15

1. Fold D along a diameter so that fD forms a semicircular region (or half disk); then any y and fz may be connected by a polygonal path that misses fC.

Section 16

1. See Fig. S6.

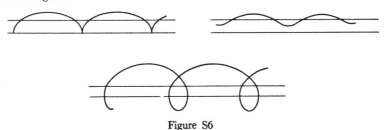

Figure S6

2. By Theorem 4.6, the composition gf is continuous. Its domain is $[a, b]$ and its range is P. Hence gf is a curve in P.

3. Similarity: $ft = a + t(b - a)$.
   Not similarity: $ft = a + t^n(b - a)$, $n \neq 1$.

Section 17

1. $90°$; $-90°$.

2. $A1$; $B2$; $C1$; $D0$; $E2$; $F0$; $G1$.

3. $A1$; $B2$; $C1$; $D1$; $E2$; $F3$; $G2$; $H1$; $I1$; $J0$.

Section 18

1. Those for which $W \neq 0$; namely, A, B, C, E, and G.

2. All but J.

SOLUTIONS 153

3. (a) *fC*: a semicircle;
   fD: all points in the region enclosed by the semicircle and the diameter.
   (b) $W = 0$. This shows that while $W \neq 0$ is a sufficient condition that a solution exists for $y = fx$, it is not a necessary condition; that is, a solution may exist even if $W = 0$.

Section 20

1. u, w, x, y, z.

2. u, 350; w, -10; x, 90; y, 180; z, 330.

Section 21

1. See Fig. 22.5 for one solution; there are many others. For example, the trivial partition consisting of the entire curve is sufficiently fine for every point of F.

2. (a) The section from a to d.
   (b) Yes; the section from f to a.
   (c) The points f and d divide the curve into two curves each of which is short relative to y.

Section 22

1. Zero.

2. (a) From a to d: $20 - 270 + (2 - 0)360 = +470$;
       from b to g: $350 - 90 + (2 - 2)360 = +260$.
   (b) 2.
   (c) For example, the ray opposite the one shown in the drawing, or any ray from y that intersects φ only twice.

Section 23

1. $A(\varphi_1, y) = +470$; $A(\varphi_2, y) = -30$; $A(\varphi \mid [t_0, t_2], y) = +440$.

Section 24

1. Any closed curve φ in P and any constant closed curve at a point y in

P are homotopic via the linear homotopy: from each point φt on φ, draw the segment from φt to y; this is the path followed by φt.

2. Let $\Phi(t,\tau) = \varphi(\tau a + (1-\tau)t)$ for $t \in [a,b]$ and $\tau \in [0,1]$. Then $\Phi(t,0) = \varphi t$ and $\Phi(t,1) = \varphi a$. The path followed by φt under the homotopy is the restriction $\varphi \mid [a,t]$ taken in reverse direction.

3. If Q is the rectangle of pairs (t,τ) such that $t \in [a,c]$ and $\tau \in [0,1]$, then Q is the union of the rectangles Q', Q'', obtained by cutting Q by the vertical line $t = b$. Let $\Phi': Q' \to P$ and $\Phi'': Q'' \to P$ be homotopies of $\varphi \mid [a,b]$ and $\varphi \mid [b,c]$ to the constant map into the point φb so that $\Phi'(b,\tau) = \varphi b = \Phi''(b,\tau)$ for all $\tau \in [0,1]$. Then Φ' and Φ'' fit together to define a mapping $\Phi: Q \to P$ where $\Phi \mid Q' = \Phi'$ and $\Phi \mid Q'' = \Phi''$.

4. Set $\Phi'(t,\tau) = \Phi(t, 1-\tau)$.

Section 25

1. The translation of φ_0 into φ_1 by the vector from the center of φ_0 to the center of φ_1 is a homotopy not meeting y.

2. Linear homotopy does not meet y; it does meet x.

3. Orientations of φ_0 and φ_1 do not correspond under the homotopy. If the orientation of φ_1, say, were reversed, they would.

Section 26

1. The map φ_0' of $[0,1]$ onto the boundary C' of the rectangle D' maps the four subintervals $[0, \frac{1}{4}]$, $[\frac{1}{4}, \frac{1}{2}]$, etc. onto the successive edges by similarities. Φ' is the linear homotopy of φ_0' into the constant closed curve at the center of D'. No other changes are needed.

Section 27

1. THEOREM: Let $f: F \to P$ be a mapping of a rectangle and its interior into a plane such that f leaves fixed each point of the periphery E of F; then the image fF contains all of F.

COROLLARY: There is no continuous mapping of a rectangle and its interior into its periphery which leaves fixed each point of the periphery.

PROOF OF THEOREM. Following the hint, let $h: P \to P$ be such that $hD = F$. Then fh maps D into P and $h^{-1}fh$ is a mapping of D into

SOLUTIONS

P. By the construction, if y is a point of the boundary C of D, then $hy \in E$ and $fhy = hy$. So $h^{-1}fhy = y$. $h^{-1}fh$ is then a mapping of a disk into a plane which leaves fixed each point of its bounding circle. By the theorem in the section, $h^{-1}fhD$ contains all of D; that is, $h^{-1}fhD \supset D$, hence $fhD \supset hD$. Since $hD = F$, this last statement says that $fF \supset F$.

2. Take f to be the projection onto the circle from y, i.e. for $x \in D - y$, fx is the point where the line through x and y meets $C - y$.

3. By radial projection from y_0 onto C.

4. (a) For example, a perpendicular projection onto the diameter.
   (b) The constant map into that fixed point.

5. There is no continuous mapping of a segment s onto its endpoints leaving fixed each endpoint, because s is connected, any continuous image of s is connected, but a set of two points is not connected.

Section 28

1. (a) The center. (b) That diameter. (c) The center.

   (d) A point on the line of centers $\frac{2}{3}$ of the radius from the center of D.

   (e) The horizontal diameter L is mapped into itself by the composite mapping f. Using a coordinate x on L with origin z we find that $fx = -\frac{1}{2}x + \frac{1}{3}r$. The fixed point occurs where $x = \frac{2}{9}r$.

   (f) Again the horizontal diameter is mapped onto itself by
   $$fx = \tfrac{1}{2}|x| - \tfrac{1}{3}r.$$
   Fixed point at $x = -\frac{2}{9}r$.

2. See answer to Exercise 4, Section 9.

Section 30

1. (a) f is the constant map discussed in the section; all vectors in the field are parallel and of the same length and orientation.

   (b) Since $fx = x$, the vector from x parallel to and of the same length and orientation as (o, fx) ends at a point y twice the distance $d(o, x)$ from o; (o, y) lies along (o, x) and is twice as long.

   (c) For every x, vx starts at x and ends at the origin.

   (d) Let the translation be described by the vector (o, c), and let z be the point such that o is the midpoint of the segment from z to c. Then the translation carries the point z into o. The vector in the

156 FIRST CONCEPTS OF TOPOLOGY

field for this point is the zero vector. All other vectors in the field "radiate" from z in the following manner. For each point x in the plane, vx is the vector issuing from x away from z on the line zx, with length equal to the distance from z to x.

(e) Let x' be the perpendicular projection of any point x on the line L of reflection. Then the vector vx goes from x to the point x'' on L which is twice as far from o as x' and on the same side of o as x'.

Section 31

1. No; vx is not defined at z, and cannot be defined there so as to be continuous.

2.

	Index	$vx = 0$	Tangents	Outward normal	Inward normal
(a)	0	none	at ends of diameter perpendicular to oz	at intersection of C with oz extended	at intersection of C with oz
(b)	1	z	none	all points of C	none
(c)	1	o	none	at ends of diameter through o and z	none
(d)	0	none	at points of contact of external tangents from o	at intersection of C with oz extended.	at intersection of C with oz
(e)	0	none	at antipodes to points of contact of external tangents from o	at intersection of C with oz extended	at intersection of C with oz
(f)	0	none	at points whose translates fx are points of contact of the tangents from o to fC	at the point which corresponds to intersection of fC with ofz extended	at the point which corresponds to intersection of fC with ofz
(g)	-1	z	at ends of the four radii making 45 degree angle with chosen diameter	at ends of chosen diameter	at ends of diameter perpendicular to chosen diameter

Section 32

1. (a) Let P' be the plane through z perpendicular to L, and C its intersection with S; then all points of C map into the intersection

SOLUTIONS   157

z of P' with L, and all pairs of diametrical points of C are antipodes having the same image.

(b) The pair $(0, r, 0)$, $(0, -r, 0)$.

2. Choose a sphere $\bar{S}$ with center at Q, and let h be the radial projection from Q of $\bar{S}$ onto S. Then h is a homeomorphism carrying ordinary antipodes on $\bar{S}$ into queer antipodes on S. Given $f: S \to P$, we get a composition $fh: \bar{S} \to P$. A pair of antipodes u, u' on $\bar{S}$ with the same image gives a pair hu, hu' of queer antipodes on S with the same image.

3. By a device similar to the above, for example, by radial projection, there is a homeomorphism from either surface to the sphere; the assertion of the problem follows.

Section 33

1. (a) The plane through the centers of the three bodies.

2. All the planes through the axis of B form a family of planes each of which divides both A and B in half. Enclose all three solids with a sufficiently large ball having the axis L of B as a diameter and center z. Each point x of the great circle E that is on the plane perpendicular to L determines a plane P_x of the family, the one perpendicular to zx. The side of P_x containing x is called its positive side. Let V_x be the volume of the part of C in the positive side of P_x, let W_x be the volume of the part of C in the negative side of P_x. Define fx to be the difference $V_x - W_x$. Then for antipodal points x, x', we have $fx' = -fx$. Since f is continuous, it must be zero somewhere on any semicircle of E.

Section 34

1. Consider the perpendicular projection ux of each vector vx into the plane tangent to the sphere at x; u is a continuous tangent field, hence is somewhere zero by the theorem; since each vx is a non-zero vector, the vector with zero tangential component is normal.

2. Radial projection maps the ellipsoid and its tangent field onto a sphere and a field tangent to it. The mapping is continuous and carries zero vectors into zero vectors.

Section 35

1. (a) Translation 4 units to the left.
   (b) Translation 2 units up.
   (c) Translation $\sqrt{2}$ units southeast.

(d) Radial expansion by a factor of 2; a similarity.
(e) 180° rotation about the origin.
(f) 90° rotation about the origin.
(g) 45° rotation about the origin followed by expansion by $\sqrt{2}$.
(h) Same as preceding followed by translation 3 units down.
(i) z^2 as described in the section, followed by the 90° rotation about the origin and by the translation NE by $2\sqrt{2}$ units.
(j) Inversion in unit circle followed by reflection in the x-axis.

Section 36

1. If $z \in E$, set $r = |z|$. Since $r > r_0$, the proof applies to the circle C of radius r, and it shows that $g(z, \tau)$ is not zero for $z \in E$ and $\tau \in [0, 1]$. When $\tau = 0$, it says that $g(z)$ is not zero. Therefore g, hence also f, has no zero in E.

2. (a) The largest of $(4 \cdot 3)^1$, $0^{1/2}$, $(4 \cdot 2)^{1/3}$, $(4 \cdot 5)^{1/4}$ is $r_0 = 12$.
   (b) $g(z) = \frac{1}{2}f(z) = z^7 + \frac{1}{2}i\, z^3 - \frac{3}{2}z$. The larger of $(7/2)^{1/4}$ and $(21/2)^{1/6}$ is $r_0 = (21/2)^{1/6}$.
   (c) $r_0 = 4\sqrt{5}$.

3. $fz = 4(z^3 - (1+i)z^2 - (2-i)z + 2i)$. The largest of $3\sqrt{2}$, $(3\sqrt{5})^{1/2}$, $(3 \cdot 2)^{1/3}$ is $r_0 = (3\sqrt{5})^{1/2}$, approximately 6.7. The absolute values of the roots 2, -1, i are 2, 1 and 1.

4. (a) The zeros are at $-2/3$ and 5, and $r_0 = 26/3$.
   (b) The zeros are at $(1 \pm i\sqrt{2})/3$ and have absolute value $\sqrt{3}/3$, and $r_0 = 4/3$.

5. Take the absolute value of the formula for $h(z)$, and use the fact that the absolute value of a sum is at most the sum of the absolute values; this gives
$$|h(z)| \leq \frac{|a_1|}{r} + \frac{|a_2|}{r^2} + \cdots + \frac{|a_n|}{r^n}.$$
Since the right side is <1, we have $|h(z)| < 1$. Now continue with the last five sentences of the proof of Theorem 36.1.

6. Taking $r = 2$ in the example, we find
$$\frac{0}{2} + \frac{|-1|}{4} + \frac{5}{8} = \frac{7}{8} < 1$$
so the disk of radius 2 contains all the zeros of the polynomial. (Compare with $r_0 = 15^{1/3}$, approximately 2.5.)

Index

Angle swept out by a curve, 88, 92
Annulus, 45
Antipodal points, 62, 116

Bound
  Lower, 44
  Upper, 43
Boundary, 79
Bounded, 37

Cartesian coordinates, 9
Chain, 135
Closed curve, 83
Closed interval, 8
Closed set, 26
Compact, 38
  Locally, 59
Complement, 10
Completeness, 34
Composition, 13
Congruence, 11
Connected, 46
  Locally, 59
Constant curve, 85, 90
Constant field, 113
Constant function, 13
Contain, 10
Continuous function, 8, 17
Contracting sequence, 34
Contraction, 11
Convex set, 50
Covering, 38
  Open, 38
Curve, 82
  Angle swept out by, 88, 92
  Closed, 83
  Constant, 85, 90
  Short, 89
Cycle, 135

Decimal expansion, 33
Dedekind cut, 37
Deformation, 99
Disconnected
  Totally, 59
Discontinuous function, 7, 18
Disk, 79
Distance, 16
Domain, 10
Duality, 27

Element, 10
Empty set, 10
Enclosing number, 133
Equivalent configurations, 58
Equivalent vectors, 112
Expansion, 11

Field
  Tangent, 123
  Vector, 111
Fixed point, 60
Function, 10
  Composition, 13
  Constant, 13
  Continuous, 8, 17
  Discontinuous, 7, 18
  Identity, 13
  Inclusion, 13
  Inverse, 13
  One-to-one, 13
  Onto, 10
Fundamental theorem of algebra, 132

INDEX

Half-open interval, 8
Homeomorphism, 54
Homology, 135
Homotopy, 98
  Linear, 99

Identity function, 13
Image, 12
  Inverse, 12
Inclusion, 13
Index of a vector field, 114
Intersection, 10
Interval
  Closed, 8
  Half-open, 8
  Open, 8
Inverse function, 13
Inverse image, 12

Linear homotopy, 99
Linking number, 134
Locally compact, 59
Locally connected, 59
Lower bound, 44

Mapping, 18
m-dimensional box, 40

Neighborhood, 17
n-tuple, 9
Number
  Algebraic, 31
  Complex, 127
  Enclosing, 133
  Linking, 134
  Pure imaginary, 127
  Rational, 31
  Real, 31, 33
  Transcendental, 31
  Winding, 82, 84, 97

One-to-one function, 13
Onto function, 10
Open covering, 38
Open interval, 8
Open set, 22

Partition, 91
  Sufficiently fine, 91
  Vertex of, 93
Perpendicular projection, 12
Polynomial
  Complex coefficients, 130
  Real coefficients, 70

Projection
  Perpendicular, 12
  Radial, 12
  Stereographic, 15

Radial projection, 12
Range, 10
Reflection, 11
Regularly contracting sequence, 34
Restriction, 13
Retract, 108
Rigid function, 20
Rotation, 11

Separation, 46
Sequence
  Contracting, 34
  Of intervals, 33
  Regularly contracting, 34
Set
  Closed, 26
  Connected, 46
  Convex, 50
  Open, 22
Short Curve, 89
Shrink to a point, 101
Similarity, 11
Space, 30
Star-shaped, 52
Stereographic projection, 15
Subset, 10
Sufficiently fine, 91

Tangent field, 123
Topological equivalence, 54
Topological property, 53
Torus, 25
Totally disconnected, 59
Translation, 11
Triangle inequality, 16

Union, 10
Upper Bound, 43

Vector field, 111
  Index of, 114
Vertex of a partition, 93

Winding number, 82, 84, 97

Zero vector, 111